Narrative Comprehension, Causality, and Coherence

Essays in Honor of Tom Trabasso

Narrative Comprehension, Causality, and Coherence

Essays in Honor of Tom Trabasso

Edited by

Susan R. Goldman
Vanderbilt University
Arthur C. Graesser
The University of Memphis
Paul van den Broek
University of Minnesota

LEA LAWRENCE ERLBAUM ASSOCIATES, PUBLISHERS
1999 Mahwah, New Jersey London

Lawrence Erlbaum Associates, Inc., Publishers
10 Industrial Avenue
Mahwah, NJ 07430

Cover design by Kathryn Houghtaling Lacey

Library of Congress Cataloging-in-Publication Data

Narrative comprehension, causality, and coherence : essays in honor of Tom Trabasso / edited by Susan R. Goldman, Arthur C. Graesser, Paul van den Broek

p. cm.
Includes bibliographical references and index.
ISBN 0-8058-3358-7 (hardcover : alk. paper).
1. Discourse analysis–Psychological aspects. 2. Causation. 3. Trabasso, Tom. I. Goldman, Susan R. II. Graesser, Arthur C. III. Broek, Paul van den.
P302.8.N37 1999
401'.41—dc21

99-19376
CIP

Books published by Lawrence Erlbaum Associates are printed on acid-free paper, and their bindings are chosen for strength and durability.

Printed in the United States of America
10 9 8 7 6 5 4 3 2 1

Contents

1

Essays in Honor of Tom Trabasso

Susan R. Goldman
Vanderbilt University
Arthur C. Graesser
University of Memphis
Paul van den Broek
University of Minnesota

Tom Trabasso is one of the founders of the field of discourse psychology. To many of us who are active in discourse psychology, it is Tom who had the intellectual vision and genuine passion that was needed to launch the field. That vision was needed to sell our ideas to experimental psychologists, cognitive scientists, and to members of the various disciplines that contribute to discourse analysis. The passion was needed for everyone to stay motivated during the confused, unsettled stage when the field was discovering and inventing itself.

Tom Trabasso's role in launching discourse psychology is a matter of public record. He was the obvious choice to be the first president of the Society for Text and Discourse (ST&D) in 1991. It was Tom who made ST&D a legal entity, while Bruce Britton prepared the bylaws and Art Graesser arranged for *Discourse Processes* to be the official journal of the Society. The first meeting of the annual ST&D conference was held within a few miles of Tom's laboratory at the University of Chicago. Tom was an associate editor of *Discourse Processes* during those formative years, sharing the duties with Roy Freedle (founding editor), Art Graesser (current editor), and Deborah Tannen. Tom Trabasso was a central figure at the meetings of the annual Winter Text Conference, held every January in

Jackson Hole, Wyoming. The Winter Text conference was organized by Bruce Britton as a forum to focus on issues of text processing, modeled after a conference held at the same location by George Sperling in the area of cognitive psychology. So every January for the past 10 years, a group of folks interested in various aspects of discourse psychology, cognition, and winter sports hold a conference in Jackson Hole. At the apex stands Tom Trabasso, along with Walter Kintsch and Chuck Perfetti, ready for a rapid, dangerous, invigorating run. The intellectual debates and provocative data are indeed as stimulating as the slopes.

Those of us who have been mentored by Tom decided to write a collection of papers to celebrate his contributions to discourse psychology. Some of us are officially his students. Some are collaborators. And some are neither, but have been deeply influenced by his ideas. The ideas in this edited volume provide a representative snapshot of research on narrative comprehension, coherence, and causality during the 1990s.

Before we describe the body of work that we have assembled in recognition of Tom's impact on discourse psychology we need to point out that this body of work represents only one of the research hats Tom has worn. Before arriving at the University of Chicago, Tom participated in the academic life of some prestigious institutions and psychology departments. Tom's formal career in psychology began at Michigan State University under the tutelage of Frank Restle, a leading figure in mathematical learning theory. Tom was awarded the PhD from Michigan in 1961 whereupon he began a National Institute of Mental Health postdoctoral fellowship at Stanford University. At Stanford, Tom worked with a budding cognitive psychologist by the name of Gordon Bower. This collaboration resulted in a lifelong friendship and sharing of intellectual ideas. At Stanford, Tom continued to pursue his research on conceptual learning and memory, interests that dated from his Master's and Doctoral theses.

Tom first joined the professorial ranks at the University of California, Los Angeles in 1963. After just 4 years, he was promoted to associate professor. He moved to Princeton University in 1969 as professor of psychology. During his time at Princeton, Tom took on the editorship of the journal *Cognitive Psychology,* becoming its second editor. Tom's research interests at Princeton eventually shifted from conceptual learning and attentional control to language processing, reasoning, and inference generation. His work examined adult performance and the development of reasoning in children. From 1976 to 1979 Tom was professor of child development at the University of Minnesota where he pursued his interests in cognitive development. While at Minnesota, Tom began collaborating with Nancy Stein. Her work on the role of story schema in narrative comprehension and memory had a substantial impact on Tom's later research on the role of causality in discourse comprehension. In 1979, Tom moved to his current home at the University of Chicago. He was instrumental in forming the department of psychology at Chicago and served as department chair for 3 years. In 1990, Tom was named the Irving B. Harris Professor of Psychology.

In addition to the variety of institutions at which Tom has held professorial positions, he has spent time in different parts of the United States and England in various visiting professor positions. Tom's mobility contributed to his impact on a broad range of scholars, as does his presence at professional meetings such as Psychonomics and Society for Research on Child Development. For various reasons, the contributors to this volume are only a small subset of all the people on whom Tom has had an impact, including some of his colleagues at the University of Chicago. Nevertheless, the volume represents the breadth and depth of Tom's influence on discourse psychology.

TOM TRABASSO'S MAJOR INTELLECTUAL CONTRIBUTIONS TO THE THEMES IN THIS FESTSCHRIFT

Although Tom's intellectual career contributes to a variety of important topics in psychology, this book focuses on his work in discourse psychology. He was instrumental in moving the area of story comprehension from an emphasis on grammatical structures to the semantic and conceptual elements that underlie those structures. Tom developed a highly influential model of the causal structure of goal-based stories (Trabasso & van den Broek, 1985). The *causal network* model has had a profound impact on the way researchers approach narrative understanding, as will be evident in a number of the chapters in this volume. As Tom pursued an indepth understanding of causal explanation for narrative events, his explorations evolved naturally to a consideration of inferential reasoning during comprehension. This path led to a collaboration with Graesser and Singer on a *Psychological Review* paper that proposed a constructionist theory of inferences and explanation-based reasoning in narrative comprehension (Graesser, Singer, & Trabasso, 1994). According to this constructionist theory, comprehenders seek to establish explanatory coherence for narrative information. Thus, even if there are explicit local connections between narrative events, comprehenders have a natural inclination to make inferences that serve explanations at deeper levels, both local and global. For example, the following segment of a story is locally coherent.

> Jimmy saw Tom's new bike. Jimmy ran home and counted his money. He called the bike store. Then he went to the local grocery store and asked for a job. He worked hard at the grocery store for several weeks. Then he had enough money.

Each of the sentences in this story segment is locally coherent. For each pair of sentences, there are at least one or two concepts that are repeated. Even though there is this local coherence, comprehenders readily make inferences about the causal or explanatory connections from sentence to sentence and across the whole segment. For example, readers will readily infer that Jimmy counted his money because he wanted a new bike like Tom's. Similarly, they infer that he went to the grocery store because he did not have enough money to buy the bike. A particularly robust finding is the propensity for readers to infer the goals that explain protagonists' actions.

The constructionist position is often contrasted with a more automatic, less strategic view of inference making in comprehension. According to this view, readers make inferences when necessary to establish coherence. Almost all of the papers in this volume address questions of what inferences are made, when inferences are made, and what representations and processing mechanisms account for these inferences.

In his most recent work, Tom has been particularly interested in the personal narratives that individuals construct and in how these narratives are diagnostic of mental health. In particular, an analysis of goal structures has been found to play a prominent role in a person's evaluations of their own experiences, emotions, attitudes, and prospects for well being. Tom has had a continuing interest, throughout his career, in the developmental course of psychological representations and skills. Consequently, some of the chapters in this volume take up the developmental theme.

OVERVIEW OF THE VOLUME

This volume begins with a short reflection (chap. 2) by two individuals whose work in discourse psychology has been powerfully influenced by Tom: Paul van den Broek and Joe Magliano. They provide some insight into the personal side of Tom. The remainder of the volume is organized into three sections. The first section deals with larger theoretical perspectives on models of discourse processing. The second section is more heavily concentrated on empirical tests and extensions of Tom's work on goal structures, causality, and coherence in narratives. The final section addresses Tom's current focal area, the use of narrative in the context of evaluative and emotional experiences.

The five chapters in the first section highlight the influence that Tom's work on causal structures has had on investigations of narrative representation and comprehension. The section begins with van den Broek and Gustafson reviewing trends in research on reading over the past 25 years (chap. 3). They reflect on the changes that have occurred in the dominant paradigms and questions driving research. Tom's work on the importance of explanation-based reasoning in constructing meaning has helped shaped the direction of this evolution. Van den Broek and Gustafson point out that theories of reading comprehension have shifted from a product to a process orientation, and more recently to an orientation that emphasizes an integrated view of process and product (van den Broek, Young, Tzeng, & Linderholm, 1999). Central to this theoretical evolution lie the challenges of dissecting the representations that readers construct and the process of constructing these representations.

The construction of meaningful narrative representations is substantially facilitated when it is easy for the reader to establish local coherence from sentence to sentence in a text. No one would quibble with this. However, when the text is locally coherent, models of comprehension differ with respect to assumptions about how much processing readers do beyond that, as discussed above. O'Brien and Myers (chap. 4) discuss and present evidence for a *resonance model* of memory-based processing. According to this view, concepts are activated through a fast-acting, passive, resonance process and are available for connection with other information currently being processed. They argue that this view may lead to a blending of two other theoretical positions, the minimalist hypothesis (McKoon & Ratcliff, 1992) and the constructionist theory (Graesser et al., 1994). Relevant to the minimalist view is the memory-based processing definition of what it means for information to be "readily available." With respect to the constructionist view, the resonance model provides a mechanism for the bottom-up, passive generation of inferences. O'Brien and Myers are particularly concerned with reactivation of concepts during reading. Empirical studies of concept reactivation provide a rich testbed for incisive contrasts between theoretical position

Chapter 5 takes a broad view on inferences in comprehension. Magliano discusses some of the theoretical and methodological issues associated with investigating inferences. Along with colleagues Trabasso and Graesser (Magliano & Graesser, 1991; Trabasso & Magliano, 1996), Magliano has argued for the importance of examining inference processes through converging methods. In this chapter, Magliano takes up three topics, corresponding to his "three-pronged method." First, he describes work that used the causal network model (Trabasso, van den Broek, & Suh, 1989) to generate theoretical predictions about where inferences ought to be made. He then presents evidence from inference elicitation methods (e.g., think-aloud protocols and question answering) of the kinds of inferences that people should make online. Finally, he provides evidence from behavioral measures about particular classes of inferences that are generated online. In general, the findings converge on the centrality of inferences that support explanation-based comprehension.

Inferences during comprehension take the reader beyond what is explicitly stated in the text (e.g., inferring Jimmy's goal of wanting a bike like Tom's). Thus, inferences are particularly important to the interpretation of the text and the construction of a more complete mental model of the situation described in a text. Although there are a number of different theories and models of the processes involved in going beyond the text, no one of them enjoys wide acceptance. In chapter 6, Graesser and Weimer-Hastings focus on predictions of the constructionist theory (Graesser et al., 1994) in accounting for the inferences that are relevant to situation model representations. They are particularly concerned with discriminating among different classes of inferences and the circumstances under which they will be made during comprehension. Inferences that are made rely on generic concepts in the reader's knowledge base. Graesser and Weimer-Hastings take up the issue of the mechanisms by which such information becomes available to readers during story comprehension.

Zwaan also takes up the issue of situation model construction in chapter 7. Zwaan builds on Tom's work that has shown how causal relations are key in situation model representations (e.g., Trabasso & Magliano, 1996; Trabasso & Sperry, 1985; Trabasso & Suh, 1993; Trabasso & van den Broek, 1985). Zwaan and his colleagues move beyond this claim by exploring other dimensions important to situation models, namely temporal, spatial, and protagonist-related connections (Zwaan, Langston, & Graesser, 1995; Zwaan & Radvansky, 1998). Zwaan describes the *event-indexing model,* which explains how these dimensions operate in concert. Zwaan discusses empirical evidence for two hypotheses derived from the event-indexing model, the *processing-load hypothesis* and the *memory-organization hypothesis.*

The second section of the book contains seven chapters that build on the causal network model to explore a range of specific phenomena. Causal constructs are examined in narrative contexts, as well as extending the use of these constructs to domain-specific learning situations in science, mathematics, and history. In chapter 8, Bower and Rinck examine the importance of characters' goals in the mental model comprehenders construct. They report data that show that psychological distances between actions and the goals they satisfy predict ease of comprehension, with shorter distances easier than longer. Furthermore, they show that memory for objects in a spatial layout is dependent on the goals of the narrative's main character.

In chapter 9, Goldman, Varma, Sharp and their colleagues build on Tom's discourse processing and developmental work to examine children's understanding of complex stories, including methods of assessing this understanding. Goldman et al. use causal network analysis to describe the causal coherence of a complex story and relate that causal structure to children's understanding. They show that understanding is more completely assessed when story production demands are removed through the use of probe questions. These results are consistent with Tom's work on task demands and their impact on estimates of children's logical reasoning competencies. In chapter 10, Boscolo and Cisotto are also concerned with children's narrative understanding. However, they examine the narratives children generate and the development of the ability to anticipate readers' inferences. They differentiate among the types of inferences that children anticipate at different ages and in two different kinds of narratives, personal narratives and invented stories.

The focus on inferences continues in Singer and Gagnon's chapter 11, but returns to issues of adult comprehension. They report work on the ability of readers to infer causal bridging inferences in expository text, a type of inference that has been studied largely in the narrative context. They first provide a theoretical analysis of causal bridging inferences in expository text, an analysis that builds on constructionist theory (Graesser et al., 1994). They examine readers' inference making through the detection of inconsistencies between causal bridging inferences and explicit text ideas. Singer and Gagnon find that readers do detect inconsistencies between inferred and explicitly presented ideas, even for expository material with which they have low familiarity. They argue that their data illustrate that readers of expository text engage in explanation-based processing, much as they do when reading narratives. However, the explanations are not necessarily accurate or particularly rich if the reader has little knowledge in the content domain of the text.

Fletcher, Lucas, and Baron take quite a different direction in chapter 12, whrere they extend causal analysis to the comprehension of mathematical proofs. They begin by noting a variety of differences between narratives and mathematical proofs. However, they argue that the path through a proof has high similarity to the kinds of causal chains that exist in narratives. They report two experiments in which they demonstrate that robust findings for narratives are also observed in the comprehension of mathematical proofs: importance of events and explanation-based processing.

The second section concludes with two chapters that explore how individuals construct causal narratives for events that have actually occurred in the real world (chap. 13) or events that could actually occur (chap. 14). So they go beyond fiction to historical fact. Britt, Perfetti, and Rouet extend causal network analysis to texts that convey information about historical events. The task for readers of history is to construct coherent, causal representations of particular historical events. They do this even when multiple documents relevant to the event do not necessarily portray the event in the same way. To account for these understanding processes, Britt et al. propose an additional level of representation, called the *documents model.* This model captures information that readers represent about the specific "source" texts. Readers extract from these texts particular information that constitutes their situation model of the historical event. Central to the situation model are the core elements of a narrative account of the historical event. These core elements define the temporal–causal structure of a narrative and are the ones that readers mark for source. They are especially likely to mark for source if different sources disagree about the central temporal–causal structure. In previous work, Britt and colleagues established the history learning was fundamentally about learning the narrative account of an event, with elements most central to the narrative learned first. Similarly, sourcing occurs first for those elements of the narrative that are most central to the temporal–causal structure.

According to some, history and courtroom arguments share common features (Collingwood, 1946; Wineburg, 1991). Voss, Wiley, and Sandak (chap. 14) point out that in both contexts there exist alternative narratives of the same event. Voss et al. argue that in the courtroom situation, as in the historical situation, narrators attempt to persuade comprehenders of the truth of their positions. In the courtroom, the comprehenders are the jury members. To form a convincing argument, attorneys strive to construct complete narratives that are causally and chronologically coherent. Voss et al. report the effects on guilty judgments of narratives that conform to these criteria of completeness and coherence, compared with narratives that are lacking along these dimensions. This work reveals how readers evaluate the plausibility of narratives about real events. The research is related to Tom's recent work on the evaluative dimension of narratives.

The final section of the book is two chapters that reflect Tom's current work on narrative. Ross, Ross, Wilson, and Smith cover the psychology of family conflict in chapter 15. Techniques of discourse and narrative analysis are used to examine the argumentation and conflict resolution strategies of young children and their parents. Disputes among siblings are the focus of the work reported in this chapter. Ross et al. rely heavily on the analysis of the goal structures of the children involved in a dispute and the emotions associated with their goal-related actions. These representational analyses allow them to expose the differences in the depictions of the same event by those involved in it.

The final chapter in the book (chap. 16) is by Stein, Sheldrick and Broaders. The productive collaboration between Tom and Nancy is evident in the issues addressed by the chapter. The specific contents of the chapter focus on appraisal processes in children and adults as they recall stressful, emotional, and conflictual events. Stein and Broaders discuss the way such events are encoded and recalled through the narratives that children and adults construct about these events. The chapter summarizes results from an impressive array of studies conducted over the past 10 years. Narrations of events are used to characterize the appraisal process. Particularly impressive is the predictive validity of these appraisals for psychological well being. This work shows how patterns of discourse may be used to understand and improve mental health.

The papers in this volume illustrate the breadth and depth of Tom's influence on the field of discourse psychology. Nevertheless, a single volume cannot reflect the extent of Tom's impact, especially considering that these chapters reflect but one set of Tom's scholarly interests. Truly his impact on cognition and its development has been positively profound. In closing this brief introduction, we wish to express our appreciation to the contributors of the volume and the editorial staff at Erlbaum who worked tirelessly with us to bring this book through the production process. We especially want to express our appreciation to Larry Erlbaum. His enthusiasm and encouragement for this project attest to his high regard for Tom and have made our efforts a uniformly positive experience.

REFERENCES

Collingwood, R. G. (1964). *Reasoning and logic.* New York, NY: Appleton-Century-Crofts.

Graesser, A. C., Singer, M., & Trabasso, T. (1994). Constructing inferences during narrative text comprehension. *Psychological Review, 101,* 371–395.

Magliano, J. P., & Graesser, A. C. (1991). A three-pronged method for studying inference generation in literary text. *Poetics, 20,* 193–232.

McKoon, G., & Ratcliff, R. (1992). Inference during reading. *Psychological Review, 99,* 440–466.

Trabasso, T., & Magliano, J. (1996). Conscious understanding during comprehension. *Discourse Processes, 22,* 255–287.

Trabasso, T., & Sperry, L. (1985). Causal relatedness and importance of story events. *Journal of Memory and Language, 24,* 595–611.

Trabasso, T., & Suh, S. Y. (1993). Understanding text: Achieving explanatory coherence through on-line inferences and mental operations in working memory. *Discourse Processes, 16,* 3–34.

Trabasso, T., & van den Broek, P. (1985). Causal thinking and the representation of narrative events. *Journal of Memory and Language, 24,* 612–630.

Trabasso, T., van den Broek, P., & Suh, S. Y. (1989). Logical necessity and transitivity of causal relations in stories. *Discourse Processes, 12,* 1–25.

van den Broek, P., Young, M., Tzeng, Y., & Linderholm, T. (1999). The landscape model of reading: Inferences and the on-line construction of a memory representation. In H. van Oostendorp & S. R. Goldman (Eds.), *The construction of mental representations during reading* (pp. 71–98). Mahwah, NJ: Lawrence Erlbaum Associates.

Wineburg, S. S. (1991). On the reading of historical texts: Notes on the breach between school and academy. *American Educational Research Journal, 28,* 495–520.

Zwaan, R. A., Langston, M. C., & Graesser, A. C. (1995). The construction of situation models in narrative comprehension: An event-indexing model. *Psychological Science, 6,* 292–297.

Zwaan, R. A., & Radvansky, G. A. (1998). Situation models in language comprehension and memory. *Psychological Bulletin, 123,* 162–185.

2

Reflections on a Mentor, Colleague, and Friend

Joseph P. Magliano
Northern Illinois University
Paul van den Broek
The University of Minnesota

This volume is dedicated to Tom Trabasso, Irving B. Harris Professor of Psychology at the University of Chicago, in celebration of his 60-something birthday and retirement from his classroom teaching career. One of the world's leading cognitive and cognitive developmental psychologists, Tom has been influential in shaping a variety of research areas through his participation in formal teaching and program development, research and research mentoring, and service to the field.

We feel particularly indebted to Tom for his mentorship of us and our work in discourse psychology. Although Tom is retiring from formal classroom teaching, this only means that he will have more time to devote to research and informal opportunities for teaching. As anyone who knows Tom will confirm, a statement such as "Tom is doing research" is a tautology. There is just no doubt that he will continue to pursue explanations of human behavior. Likewise, Tom will always be a teacher and a tutor, if not in classrooms then in other settings, such as at conferences and in personal interactions. Thus, our dedication for this volume celebrates Tom's impact on our lives. The dedication and the volume itself do not mark the end of a career, but rather seize an opportune moment to reflect on what Tom has contributed to the field and to the individuals who make their careers in this field. The chapters in this volume are testimony to Tom's influence on the area of his most recent interests, discourse processing and event comprehension. They define a case study of Tom's enormous impact on the fields of cognition and cognitive development in general.

In this brief introduction, we would like to add some reflections on Tom's impact on us as an individual, scholar, advisor, colleague, and friend. These reflections take on a personal form, but others surely will recognize many aspects that we highlight and could easily add their own examples. Both of us know Tom very well. One of us was a graduate student with Tom as advisor in the early 1980s (Paul) and the other worked with him as a postdoctoral fellow in the mid 1990s (Joe). During our respective stays at the University of Chicago, Tom set an example of intellectual curiosity and scholarly rigor for us. At the same time, we each developed strong friendships with Tom. Both our intellectual indebtedness and our friendship have continued to grow in the years since.

One of the first things that struck us both about Tom is his extraordinary desire and enthusiasm to truly understand the psychological phenomena he is investigating. Obtaining significant results or publishing another paper is irrelevant to Tom if it is not accompanied by a better grasp of the behavior. For us it meant learning that the adage "It ain't over until it's over" applies to the manner in which Tom approaches his investigations of psychological phenomena. After a hard day of working with Tom on a paper or on data analyses, success was not measured by the number of pages written or statistical analyses performed. Nor was it measured by how well and convincing our arguments were presented. Rather, it was measured by the extent to which Tom felt that he (and we in tow) had completely grasped what the data meant. A good indicator of that understanding was whether Tom would call before 8:00 the next morning to say that we needed to go over our interpretations again. Rarely did we wait for the phone call in vain! During the working sessions themselves, there often were times when we were hoping that Tom would get tired or sleepy (he never did) or that he would be satisfied with a "pretty good" argument (he never was). In the end, it set standards for us that we hope we will never lose and that we may sometimes even attain.

Working with Tom also meant learning to avoid becoming wedded to a particular theoretical framework or methodology and having it take center stage in our research. Tom continually impressed on us that theories and methodologies are tools used to understand the phenomenon under investigation, rather than ends in themselves. He would draw on cognitive psychology, developmental psychology, philosophy, linguistics, mathematics, decision-making research, anthropology, computer science, and so on, to guide his program of research. Likewise, he would use a vast array of methodologies, ranging from ones that provide speeded online measures (reading times, naming, lexical decision), to nonspeeded online measures such as think-aloud protocols, and to off-line recall and questioning techniques and occasionally questionnaires. It was the nature of the phenomenon that we were seeking to understand rather than an a priori commitment to a theory or method that dictated the nature of the research in a specific circumstance.

Tom's desire to grasp the big picture is matched by his love for empirical data. We quickly learned that Tom has a deep passion for working with data. In fact, the more complex the data, the more intrigued he becomes. Each of us can recall many occasions in which he spent the better part of a work day with Tom analyzing the think-aloud protocol of a single participant or the patterns in one subset of the data. These sessions were often agonizing and exhausting, but the fruits of our labors were immeasurable in terms of the insights we gained into how people understand texts. Tom loves to explore different ways of representing and analyzing data. He is never content with these explorations until a coherent interpretation of the data emerges. Both of us have had the experience of Tom reorganizing the manner in which we had initially presented a set of data, and in doing so, constructing a much more coherent interpretation than we initially offered.

Tom's passion for data can surface at any time and any place. When discussing some data on the way to lunch, chances were that halfway down the block Tom would stop, pull out a notepad, and proceed to graph the data to see what representation made most sense even if it was in the middle of a Chicago January with subzero temperatures (as has happened to Paul)! Or we would sit next to Tom during a conference presentation, glance over and find that he had transformed the data that were presented; he would create new graphs, tables, or bar diagrams; dimensions were switched; participant groups or conditions combined or separated. Hardly a passive listener! The result often was that Tom was able to construct a new interpretation, different from the one advanced in the presentation, sometimes to the presenter's woe and sometimes to his or her delight.

As the years of working with Tom went by, a more personal side became more and more evident to us, his tremendous generosity. His genuine care for others shows in little events. As an example, when Paul was a graduate student at the University of Chicago, he told Tom that neither his native Dutch nor his English proficiency (as evidenced by TOEFL scores and three years of academic work in the PhD program) would satisfy the University of Chicago's foreign language requirement. As a result, he would need to take the examination in German (a near-native language for most Dutch), which cost a substantial amount of money for a poor graduate student. The next day Paul found an unmarked envelope in his mailbox with $45 in it, enough to cover the registration fee for the German examination. When Joe moved from Memphis to Chicago, he was touched and amazed by the genuine concern Tom had for how well he was adjusting to life in a new city. During those initial months, Tom would take Joe to dinner, recommend places to visit, and even make sure that he took a day off every now and then to enjoy and get to know the city.

Tom's generosity is evident in the amount of time Tom spends mentoring students. Often, when teaching a student to write, Tom would sit at the computer with the student and coconstruct the initial draft of the manuscript. This kind of scaffolding requires a considerable amount of time and energy. However, the experience for a student is invaluable. He would do so with undergraduate or graduate students, with his own students as well as those who worked with others. The number of students whose theses were formed in discussions with Tom or whose first solo publication was the result of many hours of working with Tom far exceeds the official number of his advisees.

Tom's willingness to share his intellectual skills and scientific rigor is not limited to students. One need only look at the amount of time and energy that he spends with other, often junior, colleagues at conferences and via E-mail or mail, going over their theoretical ideas, the designs of their studies, or the interpretation of their data, to appreciate his dedication to assisting and fostering growth in others. Numerous colleagues who now have become established researchers in their own right can look back on the support and encouragement that they received from Tom at crucial junctures in their careers.

Tom's generosity originates in his genuine care and in his undying belief in the intellectual skills and motives of others. When he sees a seed of intellectual curiosity, he is in his element and will do whatever he can to foster that seed. Once when one of us asked him for advice on how to be a discussant for a conference symposium, he responded, "As you read each paper, ask yourself what you can learn from it. Don't try to find and emphasize weaknesses or faults, but ask yourself what this paper made you think of." More than anything, this statement reflects Tom's beliefs about how members of our scientific community should interact with one another. This sentiment is reflected throughout his interactions with others, students and colleagues alike.

In all of these respects—the standards he sets for himself in his research, his commitment to true understanding and to establishing the big picture, his generosity—Tom has made—and continues to make—an indelible impression on those who have the fortune to interact with him. To us, these qualities have made Tom an exemplary advisor, a great colleague, and a dear friend.

3

Comprehension and Memory for Texts: Three Generations of Reading Research

Paul van den Broek and Mary Gustafson
University of Minnesota

The study of reading comprehension has been a central domain of investigation by cognitive psychologists. There are good reasons for this. From a theoretical point of view, reading constitutes one of the most complex and uniquely human behaviors, and hence is an eminent domain for the study of higher human cognition in general. Basic skills (such as decoding and syntactic analysis), access to semantic and episodic memory, attention and working memory, reasoning, and knowledge of reading strategies all need to come together for reading comprehension to be successful. Thus, reading provides a window into the mind's workings and into the complexity of cognitive processes. From a practical point of view, the ability to comprehend written materials is essential for successful participation in our society. Tests of scholastic achievement, driver's license examinations, menus in restaurants, completion of tax forms, and books read for entertainment or study all require proficient reading skills. Insights gained from investigating how successful reading comprehension comes about—and why it may fail—are of great value for the development of instructional programs, design of text books, diagnosis of reading problems, and other educational practices.

Since the mid-1970s, the psychological study of reading has made enormous strides. Although advances in research are not determined by any one individual, Tom Trabasso has consistently been at the forefront of new developments in reading research, particularly with respect to the explanation-based processing used by effective readers. The context of a festschrift dedicated to Trabasso therefore provides a good opportunity to reflect on the field, its development, current issues, and future directions. In doing so we focus on the theoretical understanding of reading comprehension (for a discussion of educational implications, see van den Broek & Kremer, in press).

When taking a bird's eye view of cognitive research on reading comprehension, one can discern three *generations* of investigations (van den Broek, Young, Tzeng, & Linderholm, 1998). The focus of the first generation is on the *product* of reading: What is it that readers remember from what they read; what factors influence this memory? In the second generation, attention has shifted to the *process* of reading: What information is activated as a reader proceeds through a text; what processes occur; what factors influence these processes? More recently, the third generation aims to provide an *integration* of process and product: What are the mechanisms by which online processes lead to the construction of a stable memory representation; how does the developing memory representation in turn influence processes; what factors affect product, process, and their interaction? As with human generations, the advent of a new generation signals a shift in emphasis rather than a complete take-over: Existing generations continue to participate and to make significant contributions. Furthermore, the boundaries between generations are fluid: Individual members (in our case, theories or studies) can have characteristics of different generations even if their primary emphasis places them in one particular generation. Finally, generations reflect the spirit of their time: They depend on the accomplishments of the preceding generations as well as on changes in the external reality of their environment (in our case, technological advances, funding priorities, etc.). That said, let us consider the three generations of research more closely.

THE FIRST GENERATION: THE MENTAL REPRESENTATION OF TEXTS

A central component of successful reading comprehension is the construction of a mental representation of the text. Indeed, although people read in different circumstances and for different reasons, the desire to construct a mental representation of the text is common to most.[1] It is the nature of this representation, the product of reading, that is at the heart of research in the first generation. What are its essential features, and what factors influence its exact form and content?

Meaning Versus Surface Form

A fundamental conclusion of this research is that *the mental representation is a construction by the reader that differs from, and goes beyond, the information in the text itself.* A reader certainly can remember aspects of the exact text itself, but comprehension involves memory for the meaning of the text rather than its surface form. A meaningful representation requires that the reader recruits his or her background knowledge to interpret the textual contents. The distinction between memory for the surface structure and the meaning of a text was first introduced by Bartlett in 1932 and, reformulated by Kintsch and van Dijk (1978), has remained central to the field. A further distinction has been made between meaningful representations that are restricted to the textual elements, the *text base,* and those that include elements from background knowledge, the *situation model,* with the assumption that a reader constructs two separate meaningful representations, one of each type (Fletcher & Chrysler, 1990; Kintsch, Welsch, Schmalhofer, & Zimny, 1990; van Dijk & Kintsch, 1983). Fueled by the observation that even the most basic meaningful connections between text elements require background knowledge,[2] more recent conceptualizations modify the distinction (Coté, Goldman, & Saul, 1998; Kintsch, 1998; van den Broek, Fletcher, & Risden, 1993). Rather than being two

[1]There are a few situations in which the construction of a mental representation is not central. This is the case, for example, when the purpose for reading concerns the surface form of the text only, such as in proofreading, or when the reading process itself constitutes the sole purpose, such as some types of reading for entertainment (e.g., when passing time or reducing nervousness before boarding a plane). It should be noted that even in the latter case knowledge may be acquired, if only accidentally. (Consider, e.g., the social and moral knowledge gained from fairy tales and fables or the values and role models acquired from popular books or movies.) This function of reading may emerge in individuals or societies only after the skills to communicate/acquire knowledge have been developed (we thank Chris Espin for this suggestion).

[2]The prototypical example of a text-base connection is that of a referential or anaphoric relation. When encountering *she* or *the girl* in a text, a successful reader must resolve the anaphor by concluding that the referent refers to Mary rather than to John. Although this text-base connection is very simple and therefore may consume very few resources, it does require background knowledge, such as that *she* and *the girl* refer to female animate entities, that *Mary* but not *John* is a name for a female animate entity, and so on. Thus, except for exact repetitions, all referential connections require the application of background knowledge.

separate representations, text base and situation model in their pure forms are perhaps better considered endpoints of a dimension. Thus, a reader constructs only one meaningful representation. If it includes minimal background knowledge it can be called a text base; if it includes more extensive background knowledge it is labeled a situation model. Most often a reader's representation of a particular text will fall in between these extremes, with the exact position depending on differences in a reader's background knowledge, reading goal, and motivation, as well as on properties of the text itself.

Coherence

A second fundamental conclusion is that *a good representation is coherent.* Coherence can be defined in two distinct ways. According to the first definition, a representation is coherent if it adheres to a canonical schematic structure. For example, events and states described in narratives tend to follow certain rules of ordering. Readers follow these combinatorial rules, captured by *story grammars,* to organize events and states in their mental representation of a narrative text (e.g., Mandler & Johnson, 1977; Stein & Glenn, 1979).[3] Although less well investigated, canonical structures also exist for expository texts (e.g., Meyer, 1975). In this view, canonical schemas provide top-down constraints on the contents and form of the mental representation of a text.

According to the second definition, a representation is coherent if it includes the meaningful relations among its elements (i.e., events or states, clauses, major propositions, paragraphs, etc., depending on the particular theory). A reader can incorporate various types of relations in his or her representation, such as referential, spatial, cause-consequence, emotional, category-instance relations, and so on. As discussed later, some types of relations are included more reliably and frequently than others. Regardless of which relations are included, however, the central notion of this view is that readers identify, in a bottom-up fashion, connections among text elements and that these connections provide coherence to their mental representations (e.g., Graesser & Clark, 1985; Kintsch & van Dijk, 1978; Trabasso, Secco, & van den Broek, 1984).

[3]These rules were first noted by anthropologists (Colby, 1973; Propp, 1928) in their analyses of existing folktales.

Research on the nature of the mental representation of texts in recent years has focused primarily on the second, bottom-up/relational definition of coherence. Yet most investigators also posit that the inclusion of discourse-level knowledge in the form of schemata for different genres would likely increase the predictive power of their theories (e.g., Fletcher, van den Broek, & Arthur, 1996; Goldman, Varma, & Coté, 1996; Kintsch, 1998; van den Broek, Risden, Fletcher, & Thurlow, 1996). Including both relational and schematic constraints in a single framework has the further advantage that it allows examination of the relative importance of each. This is possible, for example, in the Recursive Transition Network model (Trabasso & Sperry, 1985; Trabasso & van den Broek, 1985), in which the story-grammatical and causal roles of text elements can be represented simultaneously.

Causal Network Representations

A coherent mental representation of a text is best captured by a network of interconnected nodes, with the nodes representing text elements and the connections indicating the semantic relations that the reader has identified between the text elements. One type of relation, causal, has been found to be particularly important in narratives as well as in various kinds of expository texts (e.g., Black & Bower, 1980; Goldman & Varnhagen, 1986; Trabasso et al., 1984; van den Broek, 1988, 1990a). This has led to the development of a causal network theory (Trabasso et al., 1984; Trabasso & Sperry, 1985; Trabasso & van den Broek, 1985; van den Broek & Lorch, 1993). It provides a set of explicit decision rules that captures the causal relations between text elements and represents them as a network (Trabasso, van den Broek, & Suh, 1989).

The most salient feature of text elements in a causal network representation is that they differ in the number of causal connections that they have to other elements. This feature has been found to be a powerful predictor of performance on memory tasks. Readers recall events with many connections more often (e.g., Goldman & Varnhagen, 1986; Graesser & Clark, 1985; Trabasso et al., 1984; Trabasso & van den Broek, 1985) and more rapidly (O'Brien & Myers, 1987), include them more often in summaries (van den Broek & Trabasso, 1986), and judge them more important (van den Broek, 1988) than events with few connections. Even children as young as 8 years (van den Broek, 1989) and—in a TV-viewing context—preschool children (van den Broek, Lorch, & Thurlow, 1996) recall events with many connections more than those with fewer connections.[4]

A second feature is that some elements are on the causal chain that connects the beginning to the end of the text. Text elements on the causal chain are recalled more often (Black & Bower, 1980; Trabasso & van den Broek, 1985), included in summaries (van den Broek & Trabasso, 1986), and judged as more important (Trabasso & van den Broek, 1985) than those that are not on the chain, even when the number of causal connections are taken into account.

Further evidence that the causal structure is a major determinant of the mental representation of texts comes from question-answering studies. There is a strong tendency for readers to provide answers that are causally connected to the queried elements (Graesser & Clark, 1985; Trabasso, van den Broek, & Liu, 1988).

Other Types of Relations: The Role of Standards for Coherence

Relations other than causal ones may also influence the mental representation of a text. For example, referential inferences allow readers to recognize the identity of persons or objects in a text. Thus, when the reader encounters a word such as *it* in a text, he or she will attempt to disambiguate the word by determining its referent. There is ample evidence that readers make such inferences (e.g., Kintsch & van Dijk, 1978; Miller & Kintsch, 1980). It has been suggested that they are a necessary, but not sufficient, step toward the representation of more semantic relations such as causal ones (Kintsch & Welsch, 1991; Trabasso et al., 1989). This would explain the finding that causal relations are a better predictor of recall and retrieval than are referential connections (e.g., O'Brien & Myers, 1987).

Other types of meaningful relations (e.g., emotional, category-instance, hierarchical theme-subtheme, spatial) exist as well. Exactly which types a reader attempts to include in his or her representation—and, within each

[4]There are systematic developmental differences, following two dimensions (e.g., Goldman & Varnhagen, 1986; van den Broek, 1996; van den Broek, Lorch, et. al., 1996). Physical, overt causal relations are detected with great reliability before children can detect internal relations such as those involving motives and goals; relations within episodes are reliably recognized before relations between episodes, and both are recognized before thematic causal relations (e.g., the moral) are.

type, how extensively[5]—depends on his or her standards for coherence (see van den Broek, Risden, & Husebye-Hartmann, 1995, for a review of factors that influence a reader's standards). Standards of referential and causal coherence seem to be almost universal—most readers in most normal reading situations appear to apply them—but at times readers apply additional standards as well (e.g., Goldman & Saul, 1990; Morrow, Bower, & Greenspan, 1989; Taylor & Tversky, 1997; Zwaan, Magliano, & Graesser, 1995).

Likewise, schematic constraints also affect memory. For example, goals have a more central place in readers' mental representations of narratives than do other types of events or states (e.g., Suh & Trabasso, 1993). Some of the centrality of goals can be attributed to the fact that they tend to have many causal relations to other elements, but even after factors such as connections are taken into account there remains a unique salience of goals (e.g., Trabasso & van den Broek, 1985; van den Broek, Lorch, et al., 1996).

Concluding Remarks on the First Generation of Reading Research

In general, findings in the first generation of research demonstrate that successful comprehension of a text involves the construction of a coherent memory representation. This representation can be depicted as a network of events, states, and facts that are connected by meaningful relations. Exactly which relations a particular reader will include depends on his or her standards for coherence, but referential and causal relations appear to be particularly fundamental.

This generation continues to investigate important issues. These include the study of possible individual and developmental differences in standards and in the ability to arrive at coherent representations, the role of the reader's goals on properties of the representation, and on the extent to which different types of text evoke different kinds of representations.

THE SECOND GENERATION: COGNITIVE PROCESSES DURING READING

The early and mid-1980s saw the birth of a new generation of research. Theoretical and methodological developments led to a new focus on the

[5]Each type of relation can be held to standards of varying stringency. For example, consider the sentence pair, *The lady gave the waiter 10 dollars. He returned to give her the change.* A reader with a modest standard for causal coherence may be satisfied by simply inferring that the lady gave too much money to the waiter, whereas a reader with a more stringent standard may be satisfied only by inferring that she must have consumed something, that the scene takes place in a restaurant, and so forth.

cognitive processes that take place during reading. On the theoretical side, there was a growing awareness in cognitive psychology that limitations in human attentional resources seriously constrain cognitive processes. This concept was introduced in reading research by the publication of a theoretical account that highlighted the role of attentional limitations in reading in particular (Just & Carpenter, 1980). On the methodological side, the development of new techniques allowed researchers to investigate and measure online activities (e.g., eye-tracking techniques, computer-directed speeded tasks, etc.) Thus, the aim of this generation is to determine what cognitive processes take place as the reader proceeds through a text, and what factors influence these processes.

Attention Allocation and Online Activations

A fundamental notion in this generation of research is that the reader's attentional or working-memory resources are limited and, hence, that at any point during reading he or she can only attend to a small subset of all elements that are relevant to the text. As a consequence, the reader must wisely select which elements to attend to if he or she is to be successful in comprehending a text.

As a result of the *allocation of attention,* different elements will be activated at different points in the text. A central question for this generation of research is what elements or concepts are activated at a particular point in reading (Fletcher & Bloom, 1988; van den Broek, 1990b). One obvious source of activated elements is the text itself: At each point in reading, the elements in the currently read text segment will be activated. There is also general agreement that elements that are associated to those currently read—either through the reader's semantic knowledge or through the current reading—will be activated (e.g., Kintsch, 1988; van den Broek, 1990b).

Inference Generation During Reading

The third possible source of activated elements, inferences, is the focus of much debate. Indeed, the majority of second-generation research is focused on determining which inferences readers make and which they do not make (Graesser, Singer, & Trabasso, 1994; McKoon & Ratcliff, 1992; Singer, Graesser, & Trabasso, 1994; see Singer, 1994, and van den Broek, 1994, for reviews).

As the findings of the first generation of studies show, successful comprehension involves the construction of an interconnected, coherent representation of the text. To do so, the reader must infer relations between the text elements that he or she encounters. It seems likely that many, if not all, of these inferences are made during the reading process itself. Among the candidate inferences are those concerning referential, causal, spatial, instrument, and thematic relations.

There is considerable evidence that readers routinely activate information that assists them in generating referential (e.g., Corbett & Chang, 1983; Gernsbacher, 1989; Haviland & Clark, 1974; McKoon & Ratcliff, 1980; O'Brien, Duffy, & Myers, 1986) and causal inferences (e.g., Bloom, Fletcher, van den Broek, Reitz, & Shapiro, 1990; Casteel, 1993; Singer, Halldorson, Lear, & Andrusiak, 1992; Trabasso & Suh, 1993; van den Broek & Thurlow, 1990). The evidence on other candidate inferences is mixed (e.g., on spatial inferences, compare McKoon & Ratcliff, 1992, with Glenberg & Langston, 1992, and Morrow et al., 1989).

In recent years, the focus of this generation has shifted from the all-or-none question of whether a particular type of inference is or is not generated to the more refined question of under what circumstances that type of inference is generated. This shift in focus has led to the development of several models aimed at capturing the interaction between attentional resources and inference making. Although the models differ in their details and their specific emphases, they share several important properties. These properties are best explained in the context of a specific model. We illustrate them by reference to the *Causal Inference Maker* model (van den Broek, 1990b, 1994).

The Causal Inference Maker

As with other models,[6] the Causal Inference Maker model (van den Broek, 1990b, 1994) is based on the assumption that the reader's attentional resources are limited and that activation is selective. In addition to the contents of the currently read sentence and its associations, inferences are the major determinant of online activations. Moreover (and in agreement

[6]Examples of other models are the *Current State Strategy* (Fletcher & Bloom, 1988), the *Construction-Integration* model (Kintsch, 1988), the *Structure Building Framework* (Gernsbacher, 1990), minimalist theories (McKoon & Ratcliff, 1992), constructionist theories (Graesser, Singer, & Trabasso, 1994; Singer, Graesser, & Trabasso, 1994), and the *3CI* model (Goldman & Varma, 1995). (See also van Oostendorp & Goldman, 1998, for examples of various models.)

with the other models), inferences occur primarily as readers attempt to maintain coherence for each event or state that they encounter in the text. To do so, they may access episodic memory, reactivating information from their memory for the prior text, possibly looking back if the prior text is still available, (Coté et al., 1998), or they may access semantic memory, retrieving background knowledge.

The fact that inferences are driven by an attempt to maintain coherence makes it necessary to adopt a functional view of inference making: Whether a particular inference is generated depends on its role in maintaining coherence. Thus, standards for coherence, which play such an important role in the product of reading, also are an important factor in inference generation.

The Causal Inference Maker model predicts that inferences are generated when they are needed to provide causal coherence to the text. There is considerable support for these predictions. Readers slow down when an adequate causal relation is missing (Bloom et al., 1990; van den Broek & Thurlow, 1990), they reactivate missing antecedents from prior text (Myers, Shinjo, & Duffy, 1987; Suh & Trabasso, 1993; van den Broek, Rohleder, & Narváez, 1996; van den Broek & Thurlow, 1990), or activate them from background knowledge (Singer et al., 1992; van den Broek, Rohleder, & Narváez, 1996; van den Broek & Thurlow, 1990).

The role that inferences play in maintaining causal—or other—coherence may also be responsible for the ambiguous findings with respect to other types of inferences. For example, although the evidence on spatial inferences is mixed, closer examination suggests that whether spatial inferences are made or not depends in part on their causal function: If the potential spatial inference provides causal information that is important to comprehension, then the inference will be made (van den Broek, Zwaan, Sundermeier, & Gaddy, 1997). Again, the categorical question of whether a particular type of inference is made or not during reading is too simplistic: The likelihood that spatial, emotional, or temporal relations are inferred depends on their function in establishing the level of coherence that the reader pursues.

Concluding Remarks on the Second Generation of Reading Research

In general, the findings of the second generation of research demonstrate that readers have limited attentional resources and hence are selective in allocating their attention. Furthermore, inference generation is functional: Inferences tend to be generated when they are needed to provide coherence. Both causal and referential inferences are the most prevalent, but other types of inferences can be made, especially when they provide causal or referential coherence as well or when the reader has special goals or standards for coherence in reading.

In all, whether a particular inference is generated in a particular instance depends on the interaction of several factors. The most important of these factors are the constraints that the text provides for the hypothesized inference, the reader's background knowledge relevant to the inference, the reader's attentional resources, and the reader's criteria for coherence. The first three factors concern the ease with which the target inference can be made; the fourth concerns the relevance for comprehension of the inference. Together they determine the ebb and flow of inference-based activations during reading.

THE THIRD GENERATION: FROM INFERENTIAL PROCESSES TO MEMORY REPRESENTATION

The first and second generations of research have greatly advanced our understanding of the nature of the memory representation and of the cognitive processes during reading, respectively. The aim of the third generation of research is to combine and extend these accomplishments by providing an integrated account that describes how online activations lead to a coherent memory representation.

The Reading Process: Patterns of Fluctuating Activations

Central to models in the third generation is the idea that reading is a continuous and dynamic shifting of activations. At each point during the reading process, many textual or inferred elements are activated. As the next reading cycle (e.g., the next sentence) is being read, this pattern or vector changes as a result of the new information and the ensuing inferential and associative processes. This notion is based on the theories developed in the second generation, but the third generation places greater emphasis on the dynamic aspect of reading. Most studies in the second generation are aimed at demonstrating that a certain inference is made under certain specified circumstances. To do so, researchers select one or two locations in their experimental texts where the to-be-tested theories make clear and strong predictions about the activation of a single text element or inference. The experimental results are then taken to indicate that the predicted element or inference is or is not made at that particular point. In contrast, research in the third generation is aimed at keeping track of the activations of multiple textual elements and inferences throughout the reading process (Goldman & Varma, 1995; Kintsch, 1988; Langston & Trabasso, 1998; van den Broek, Risden, et al., 1996; van den Broek et al., 1998). Thus, the reading process is considered to consist of patterns of continuously fluctuating activations (van den Broek, Risden, et al., 1996; van den Broek et al., 1998). As a result, the online component of the third generation differs from the second-generation models in several respects.

Degrees of Activation. The emphasis on the dynamic behavior of multiple elements has led to the realization that text elements and inferences can be activated to different degrees rather than in an all-or-none fashion (Goldman & Varma, 1995; Kintsch, 1988; van den Broek, Risden, et al., 1996; van den Broek et al., 1998). The assumption of all-or-none activations, useful when testing the activation of a single, strongly supported inference,[7] becomes implausible when one needs to simultaneously represent text elements and inferences that are supported by the text to different degrees, that may be activated through different mechanisms (e.g., association or coherence-based inference), that were activated recently although not in the current sentence, and so on. In third-generation models, activation fluctuates along a continuous dimension.

Attentional Limitations: Pool of Activation Versus Fixed Buffer. A reader's attentional resources are considered to represent a pool of activation that can be distributed over many or few text elements and inferences rather than a fixed buffer with a limited number of slots. This view follows in part from researchers' recognition that activation of text elements or inferences varies along a continuum. The imposition of an artificial dichotomy, although useful in the initial development of reading theories, ultimately is unnecessary and, moreover, leads to a considerable loss of information.

[7]Even when investigating single, strongly supported inferences, one need not assume all-or-none activation. Indeed, several second-generation studies have hinted at the existence of different levels of activation (e.g., McKoon & Ratcliff, 1989; O'Brien & Myers, 1987).

Sources of Activation. All third-generation models include both the currently read text and carry-over from the preceding cycle as sources of activation. The models differ, however, with respect to other text elements or inferences that they deem activated at each reading cycle. In general, the differences resemble those among theories of the second generation. Some models also allow reactivations of information from the developing episodic representation of the text, whereas others only consider semantic knowledge as an additional source of activation. Likewise, the models differ in the criteria for coherence they assume readers to have: Some assume that readers attempt to maintain referential coherence; others emphasize the importance of causal coherence. The *Landscape* model, outlined later, includes both episodic and semantic memory as a source of input, and both referential and causal coherence as criteria that readers normally employ.

Building a Memory Representation

As the reader proceeds through the text, he or she constructs an episodic memory representation. New text and inferred elements are added to the existing representation as are new relations between the elements; existing memory nodes and relations are strengthened by repeated exposure. In general the memory representation is depicted as a network, similar to those described in the first generation of research. The third generation thus illuminates how readers establish an increasingly complete and stable representation of the text and inferences made during reading.

Updating the Memory Representation. Readers update their representation only when elements become activated. Thus, a new element is added or the strength of an existing node increased only when it is activated (current text, carry-over, prior text, background knowledge). Likewise, a relation between elements is added or modified only if the two elements are activated simultaneously. Changes in the strengths of nodes or relations can be linear or asymptotic. In the former case, learning (i.e., the amount of change in the representation) is a function solely of the incoming information. In the latter case, it is a function of both the incoming information and the node and relation strengths in the representation as it has been constructed over prior cycles. The latter case has interesting implications: It implies, for example, that the first occurrence of a text element or inference (or a pair of elements/inferences) brings about a bigger change in the episodic representation than subsequent occurrences.

Reciprocal Relation Between Representation and Activation Patterns. The sequence of activation patterns that occur during reading thus gradually give rise to a complete representation of what has been read. Conversely, the developing episodic representation may influence subsequent activation patterns. This may occur, for example, when reading the current sentence reactivates information encountered in a preceding cycle (either through passive spread of activation or through an active effort to comprehend). It seems likely that in such situations elements and inferences that have been connected to the reactivated information in the episodic representation also become activated and part of the process of comprehending the current sentence. In this view, the reader's interpretation of a new sentence is in part determined by his or her interpretation of the entire preceding text. Indeed, activation of one member of a pair of concepts that have been connected earlier has been found to reactivate the other member (e.g., McKoon & Ratcliff, 1980; O'Brien & Albrecht, 1992). Only one current model of reading, the Landscape model, includes the developing episodic representation as a source of activation.

Computational Models

All models in the third generation use computational approaches to capture both the complexities of activations at each point during reading and the gradual emergence of a memory representation. These computer simulations allow researchers to treat incoming information as an input vector, and the developing episodic representation as a matrix of strengths. A major advantage of computational approaches is that they can capture the full complexity of information management that occurs during reading.

The Landscape Model

The Landscape model (van den Broek, Risden, et al., 1996; van den Broek et al., 1998) illustrates the features of third-generation models described in the preceding sections. Central to this model is the notion that elements that are either explicitly mentioned in the text or that are inferred fluctuate in their activation over the course of reading. These fluctuations form a landscape of activation. Activations come from (a) the contents of current input, (b) carry-over from the preceding input cycle, (c) coherence-based inferences (through reactivations from the prior text or retrieval from semantic knowledge), and (d) spread of activation to the current episodic memory representation. Different criteria for coherence can be implemented in the Landscape model: Based on the findings in the first and second generations of research, the current implementations of the model assume that readers routinely maintain referential and causal coherence.

The representational strengths of elements and relations between elements are updated in an asymptotic fashion, as described in the preceding section. (For a full description of the Landscape model, see van den Broek et al., 1998.) Moreover, the strength of the relation between two elements depends in part on the connections that each element has to other elements. Thus, elements compete with each other over the power to predict other elements.

The Landscape model predicts both online activations and off-line representations. For example, the model was found to account for more than 50% of the variance in readers' online activations (van den Broek, Risden, et al., 1996). At the same time, it accounted for 64% of the variance in readers' memory and a substantial portion of the order in which readers recalled the texts (98% of first mentions, 55% of subsequent mentions). Thus, it effectively predicts the cognitive processes during reading as well as the resulting memory representation.

Concluding Remarks on the Third Generation of Reading Research

In general, the models in the third generation of reading research build on and extend the conclusions of the first and second generations by integrating the reading process and the product of comprehension in a unified theoretical framework. In doing so, some of the central ideas in reading research have been refined. Attentional resources form a pool of activation that is distributed over multiple elements, with some elements receiving more and others less activation. Similarly, connections in the mental representation are not all or none but can vary in their strengths. Thus, the models present a distinctly dynamic and fluid picture of the processes and product of reading comprehension.

Third-generation models can be used to implement and contrast different reading theories. For example, in its current form the Landscape model assumes that readers attempt to maintain referential and causal coherence. Other criteria of coherence can also be used, yielding different activation patterns during comprehension and different representations after reading. In this way, researchers can contrast minimalist models (in which only connections between adjacent sentences are included), explanation-based models (in which connections to nonadjacent sentences are also included if they are necessary for causal coherence), and constructionist models (in which all possible causal connections are included). Results of such comparisons indicate that explanation-based models best capture readers' on-line and off-line behaviors (Risden, 1996).[8]

CONCLUDING REMARKS

Reading comprehension is successful if the cognitive processes that take place as the reader proceeds through the text result in a coherent memory representation, which can be accessed at a later point in time to retell the story, answer questions, apply to a real-life situation, and so on. The three generations of research described in this chapter provide a rich description of the process and product of reading comprehension. With the arrival of the third generation, efforts to construct a unified framework for simultaneously considering process and product of comprehension have begun. As this generation matures, we can expect modifications to the current models, for example, to capture the exact role of background knowledge, and to accommodate individual and developmental differences in knowledge and attentional resources. Moreover, the models in the third generation may spur renewed investigations in the first two generations, for instance, considering variations in strengths of relations in the mental representation of a text after reading or degrees of activation during reading.

Classifying reading research into these three generations illustrates both the continuities and the diversions of cognitive research. It shows, for instance, that research trends emerge independently, albeit roughly simultaneously, in individual researchers. This may be due partly to the fact that the arrival of a new generation is facilitated by technological advances; it is also caused, however, by a critical mass of findings in the existing generations that enable the conceptual leap to a new generation.

What will future generations of research in reading look like? It is apparent that the third generation will fuel continued attention to the access of semantic and episodic memory representations and to the dynamic and fluid aspects of activations during reading. At the most detailed level, this attention is likely to materialize in direct investigation of the neurological

---

[8]Interestingly, the second-best model of online behaviors was the minimalist model, whereas the second-best model for off-line behaviors was the constructionist model. This pattern suggests that some connections are constructed or strengthened after reading has been completed (Risden, 1996).

activities involved in comprehension and memory for texts. With the proliferation of neural imaging techniques, the technological basis for such a new trend (generation, even?) is now available.

REFERENCES

Bartlett, F. C. (1932). *Remembering: A study in experimental and social psychology.* Cambridge, MA: Cambridge University Press.

Black, J. B., & Bower, G. H. (1980). Story understanding as problem solving. *Poetics, 9,* 223–250.

Bloom, C. P., Fletcher, C. R., van den Broek, P., Reitz, L., & Shapiro, B. P. (1990). An on-line assessment of causal reasoning during comprehension. *Memory & Cognition, 18,* 65–71.

Casteel, M. A. (1993). Effects of inference necessity and reading goal on children's inferential generation. *Developmental Psychology, 29*(2), 346–357.

Colby, B. N. (1973). A partial grammar of Eskimo folktales. *American Anthropologist, 75,* 645–662.

Corbett, A. T., & Chang, F. R. (1983). Pronoun disambiguation: Accessing potential antecedents. *Memory & Cognition, 11,* 283–294.

Coté, N. A., Goldman, S. R., & Saul, E. U. (1998). Students making sense of informational text: Relations between processing and representation. *Discourse Processes, 25*(1), 1–53.

Fletcher, C. R., & Bloom, C. P. (1988). Causal reasoning in the comprehension of simple narrative texts. *Journal of Memory and Language, 27,* 236–244.

Fletcher, C. R., & Chrysler, S. T. (1990). Surface forms, textbases, and situation models: Recognition memory for three types of textual information. *Discourse Processes, 13,* 175–190.

Fletcher, C. R., van den Broek, P., & Arthur, E. J. (1996). A model of narrative comprehension and recall. In B. K. Britton & A. C. Graesser (Eds.), *Models of understanding text* (pp. 141–163). Hillsdale, NJ: Lawrence Erlbaum Associates.

Gernsbacher, M. A. (1989). Mechanisms that improve referential access. *Cognition, 32,* 99–156.

Gernsbacher, M. A. (1990). *Language comprehension as structure building.* Hillsdale, NJ: Lawrence Erlbaum Associates.

Glenberg, A. M., & Langston, W. E. (1992). Comprehension of illustrated text: Pictures help to build mental models. *Journal of Memory and Language, 31,* 129–151.

Goldman, S. R., & Saul, E. U. (1990). Flexibility in text processing: A strategy competition model. *Learning and Individual Differences, 2,* 181–219.

Goldman, S. R., & Varma, S. (1995). CAPping the construction-integration model of discourse comprehension. In C. Weaver, S. Mannes, & C. Fletcher (Eds.), *Discourse comprehension: Models of processing revisited* (pp. 337–358). Hillsdale, NJ: Lawrence Erlbaum Associates.

Goldman, S. R., Varma, S., & Coté, N. (1996). Extending capacity-constrained construction-integration: Toward "smarter" and flexible models of text comprehension. In B. K. Britton & A. C. Graesser (Eds.), *Models of understanding text* (pp. 73–114). Mahwah, NJ: Lawrence Erlbaum Associates.

Goldman, S. R., & Varnhagen, C. K. (1986). Memory for embedded and sequential story structures. *Journal of Memory and Language, 25,* 401–418.

Graesser, A. C., & Clark, L. F. (1985). *The structures and procedures of implicit knowledge.* Norwood, NJ: Ablex.

Graesser, A. C., Singer, M. & Trabasso, T. (1994) Constructing inferences during narrative text comprehension. *Psychological Review, 101,* 371–395.

Haviland, S. E., & Clark, H. H. (1974). What's new? Acquiring new information as a process in comprehension. *Journal of Verbal Learning and Verbal Behavior, 13,* 512–521.

Just, M. A., & Carpenter, P. A. (1980). A theory of reading: From eye fixations to comprehension. *Psychological Review, 87,* 329–354.

Kintsch, W. (1988). The role of knowledge in discourse comprehension: A construction-integration model. *Psychological Review, 95,* 163–182.

Kintsch, W. (1998). *Comprehension: A paradigm for cognition.* New York: Cambridge University Press.

Kintsch, W. A., & van Dijk, T. A. (1978). Toward a model of text comprehension and production. *Psychological Review, 85,* 363–394.

Kintsch, W., & Welsch, D. M. (1991). The construction-integration model: A framework for studying memory for text. In W. Hockley & S. Lewandowsky (Eds.), *Relating theory and data: Essays on human memory in honor of Bennet B. Murdock* (pp. 367–385). Hillsdale, NJ: Lawrence Erlbaum Associates.

Kintsch, W., Welsch, D., Schmalhofer, F., & Zimny, S. (1990). Sentence memory: A theoretical analysis. *Journal of Memory and Language, 29,* 133–159.

Langston, M., & Trabasso, T. (1998). Identifying causal connections and modeling integration of narrative discourse. In H. van Oostendorp & S. R. Goldman (Eds.), *The construction of mental representations during reading* (pp. 29–69). Mahwah, NJ: Lawrence Erlbaum Associates.

Mandler, J. M., & Johnson, N. S. (1977). Remembrance of things parsed: Story structure and recall. *Cognitive Psychology, 9,* 111–151.

McKoon, G. & Ratcliff, R. (1980). The comprehension processes and memory structures involved in anaphoric reference. *Journal of Verbal Learning and Verbal Behavior, 19,* 668–682.

McKoon, G., & Ratcliff, R. (1989). Semantic associations and elaborative inference. *Journal of Experimental Psychology: Learning, Memory, and Cognition, 15,* 326–338.

McKoon, G., & Ratcliff, R. (1992). Inferences during reading. *Psychological Review, 99,* 440–466.

Meyer, B. J. F. (1975). *The organization of prose and its effects on memory.* Amsterdam: North-Holland.

Miller, J. R., & Kintsch, W. (1980). Readability and recall of short prose passages: A theoretical analysis. *Journal of Experimental Psychology: Human Learning and Memory, 6,* 335–354.

Morrow, D. G., Bower, G. H., & Greenspan, S. E. (1989). Updating situation models during narrative comprehension. *Journal of Memory and Language, 24,* 304–319.

Myers, J. L., Shinjo, M., & Duffy, S. A. (1987). Degree of causal relatedness and memory. *Journal of Memory and Language, 26,* 453–465.

O'Brien, E. J., & Albrecht, J. E. (1992). Comprehension strategies in the development of a mental model. *Journal of Experimental Psychology: Learning, Memory, and Cognition, 18,* 777–784.

O'Brien, E. J., Duffy, S. A., & Myers, J. L. (1986). Anaphoric inference during reading. *Journal of Experimental Psychology: Learning, Memory, and Cognition, 12,* 346–352.

O'Brien, E. J., & Myers, J. L. (1987). The role of causal connections in the retrieval of text. *Memory & Cognition, 15,* 419–427.

Propp, V. (1928). *Morphology of the folktale.* Austin: University of Texas Press.

Risden, K. (1996). *Causal inference in narrative text comprehension.* Unpublished doctoral dissertation, University of Minnesota, Minneapolis.

Singer, M. (1994). Discourse inference processes. In M. A. Gernsbacher (Ed.), *Handbook of psycholinguistics* (pp. 479–516). New York: Academic Press.

Singer, M., Graesser, A. C., & Trabasso, T. (1994). Minimal or global inference during reading. *Journal of Memory and Language, 33,* 421–441.

Singer, M., Halldorson, M., Lear, J. C., & Andrusiak, P. (1992). Validation of causal bridging inferences in discourse understanding. *Journal of Memory and Language, 31,* 507–524.

Stein, N. L., & Glenn, C. G. (1979). An analysis of story comprehension in elementary school children. In R. O. Freedle (Ed.), *New directions in discourse processing* (Vol. 2, pp. 53–120). Norwood, NJ: Ablex.

Suh, S. Y., & Trabasso, T. (1993). Inferences during reading: Converging evidence from discourse analysis, talk-aloud protocols and recognition priming. *Journal of Memory and Language, 32,* 279–300.

Taylor, H. A., & Tversky, B. (1997). Indexing events in memory: Evidence of index dominance. *Memory, 5*(4), 509–542.

Trabasso, T., Secco, T., & van den Broek, P. W. (1984). Causal cohesion and story coherence. In H. Mandl, N. L. Stein, & T. Trabasso (Eds.), *Learning and comprehension of text* (pp. 83–111). Hillsdale, NJ: Lawrence Erlbaum Associates.

Trabasso, T., & Sperry, L. L. (1985). Causal relatedness and importance of story events. *Journal of Memory and Language, 24,* 595–611.

Trabasso, T., & Suh, S. (1993). Understanding text: achieving explanatory coherence through on-line inferences and mental operations in working memory. *Discourse Processes, 16,* 3–34.

Trabasso, T., & van den Broek, P. (1985). Causal thinking and the representation of narrative events. *Journal of Memory and Language, 24,* 612–630.

Trabasso, T., van den Broek, P. W., & Liu, L. (1988). A model for generating questions that assess and promote comprehension. *Questioning Exchange, 2,* 25–38.

Trabasso, T., van den Broek, P., & Suh, S. (1989). Logical necessity and transitivity of causal relations in the representation of stories. *Discourse Processes, 12,* 1–25.

van den Broek, P. (1988). The effects of causal relations and hierarchical position on the importance of story statements. *Journal of Memory and Language, 27,* 1–22.

van den Broek, P. W. (1989). Causal reasoning and inference making in judging the importance of story statements. *Child Development, 60,* 286–297.

van den Broek, P. (1990a). The causal inference maker: Towards a process model of inference generation in text comprehension. In D. A. Balota, G. B. Flores d'Arcais, & K. Rayner (Eds.), *Comprehension processes in reading* (pp. 423–445). Hillsdale, NJ: Lawrence Erlbaum Associates.

van den Broek, P. W. (1990b). Causal inferences in the comprehension of narrative texts. In A. C. Graesser & G. H. Bower (Eds.), *Psychology of learning and motivation: Inferences and text comprehension* (Vol. 25, pp. 175–194). San Diego: Academic Press.

van den Broek, P. (1994). Comprehension and memory of narrative texts: Inferences and coherence. In M. A. Gernsbacher (Ed.), *Handbook of psycholinguistics* (pp. 539–588). New York: Academic Press.

van den Broek, P. (1996). Discovering the cement of the universe: The development of event comprehension from childhood to adulthood. In P. W. van den Broek, P. J. Bauer, & T.

Bourg (Eds.), *Developmental spans in event comprehension and representation* (pp. 321–342). Mahwah, NJ: Lawrence Erlbaum Associates.

van den Broek, P., Fletcher, C. R., & Risden, K. (1993). Investigations of inferential processes in reading: A theoretical and methodological integration. *Discourse Processes, 16,* 169–180.

van den Broek, P., & Kremer, K. (in press). The mind in action: What it means to comprehend during reading. In B. M. Taylor, P. van den Broek, & M. Graves (Eds.), *Reading for meaning.* New York: Teachers College Press.

van den Broek, P., & Lorch, R. F., Jr. (1993). Network representations of causal relations in memory for narrative texts: Evidence from primed recognition. *Discourse Processes, 16,* 75–98.

van den Broek, P. W., Lorch, E. P., & Thurlow, R. (1996). Children's and adults' memory for television stories: The role of causal factors, story-grammar categories and hierarchical level. *Child Development, 67*(6), 3010–3028.

van den Broek, P., Risden, K., Fletcher, C. R., & Thurlow, R. (1996). A "landscape" view of reading: Fluctuating patterns of activation and the construction of a stable memory representation. In B. K. Britton & A. C. Graesser (Eds.), *Models of understanding text* (pp. 165–187). Mahwah, NJ: Lawrence Erlbaum Associates.

van den Broek, P., Risden, K., & Husebye-Hartmann, E. (1995). The role of readers' standards for coherence in the generation of inferences during reading. In R. F. Lorch, Jr. & E. J. O'Brien (Eds.), *Sources of coherence in text comprehension* (pp. 353–373). Hillsdale, NJ: Lawrence Erlbaum Associates.

van den Broek, P. W., Rohleder, L., & Narváez, D. (1996). Causal inferences in the comprehension of literary texts. In R. J. Kreuz & M. S. McNealy (Eds.), *Empirical approaches to literature and aesthetics* (pp. 179–200). Norwood, NJ: Ablex.

van den Broek, P., & Thurlow, R. (1990, November). *Reinstatements and elaborative inferences during the reading of narratives.* Paper presented at the 31st annual meeting of the Psychonomic Society, New Orleans.

van den Broek, P., & Trabasso, T. (1986). Causal networks versus goal hierarchies in summarizing texts. *Discourse Processes, 9,* 1–15.

van den Broek, P., Young, M., Tzeng, Y., & Linderholm, T. (1998). The Landscape model of reading: Inferences and the online construction of a memory representation. In H. van Oostendorp & S. R. Goldman (Eds.), *The construction of mental representations during reading* (pp. 71–98). Mahwah, NJ: Lawrence Erlbaum Associates.

van den Broek, P., Zwaan, R. A., Sundermeier, B., & Gaddy, M. (1997, November). *Activation and representation of spatial inferences in reading: The role of causality.* Paper presented at the 38th annual meeting of the Psychonomic Society, Philadelphia.

van Dijk, T. A., & Kintsch, W. (1983). *Strategic discourse comprehension.* New York: Academic Press.

van Oostendorp, H., & Goldman, S. R. (Eds.). (1998). *The construction of mental representations during reading.* Mahwah, NJ: Lawrence Erlbaum Associates.

Zwaan, R. A., Magliano, J., & Graesser, A. C. (1995). Dimensions of situation model construction in narrative comprehension: An event-indexing model. *Psychological Science, 6,* 292–297.

4

Text Comprehension: A View From the Bottom Up

Edward J. O'Brien
University of New Hampshire
Jerome L. Myers
University of Massachusetts

A basic principle in writing is that ideas should flow easily, one from another. In this way, as each new idea is introduced, it is easily integrated with the information that immediately precedes it. That is, the text should always be locally coherent. Models of text comprehension agree that readers make use of the information in working memory when integrating each new sentence into the text representation. There is less agreement about when a sentence is integrated with information that is not currently in working memory. Models of comprehension differ in the extent to which it is assumed that readers will access relevant distant information when the text is locally coherent (e.g., Fletcher & Bloom, 1988; Glenberg & Langston, 1992; Graesser, Singer, & Trabasso, 1994; Kintsch, 1988; Kintsch & van Dijk, 1978; McKoon & Ratcliff, 1992; Singer, Graesser, & Trabasso, 1994).

Two contrasting positions with respect to this issue are the "minimalist" and "constructionist" views of text processing . The critical difference in their positions is the extent to which information from earlier portions of the text, as well as from general world knowledge, is available to the reader, and how that information becomes available. According to a strong version of the minimalist view, the amount of information that is readily available to the reader is quite limited. Readers focus primarily on maintaining local coherence; as long as current information can be readily integrated with information already in working memory (i.e., the text is locally coherent), no connections are made with relevant information from earlier in the text, nor is the activation of additional information from general world knowledge required. This view of comprehension is considered minimalist because the reader's role in gaining access to information necessary for integration is passive; the reader "works" only with what is readily available unless a coherence break forces a search for additional information.

In contrast, according to a strong version of the constructionist view, readers continually search after meaning. Driven by a "need to know" node (e.g., Singer et al., 1994), readers always seek explanations and attempt to fully integrate current information with all relevant prior information. This view is considered constructionist because the reader is actively engaged in searching memory for relevant information, making every effort to ensure full integration at both local and global levels. The representation that would result from such a process has been best captured by the early work of Trabasso and his colleagues (e.g., Trabasso & Sperry, 1985; Trabasso & van den Broek, 1985) in which all possible causal connections were assumed to be made.

In the extreme, neither view can be correct. Contrary to a strong version of the minimalist view, there have been several demonstrations that information presented earlier in a text may be reactivated even when a text is locally coherent, and there is no need for a conscious search for information related to current input (e.g., Albrecht & Myers, 1995, 1998; Albrecht & O'Brien, 1993; Cook, Halleran, & O'Brien, 1998; McKoon, Gerrig, & Greene, 1996; Myers, O'Brien, Albrecht, & Mason, 1994; O'Brien, Rizzella, Albrecht, & Halleran, 1998). And, although there is no direct evidence to refute a strong version of the constructionist view, limited capacity constraints dictate that it cannot be correct (e.g., Fletcher & Bloom, 1988). Readers cannot possibly find and make every possible connection.[1] If they did, as text length increased, reading rate would have to continually slow because the reader would have to actively search increasing amounts of information to ensure that full comprehension was maintained. Indeed, research using talk-aloud protocols suggests that although readers may notice many potential connections, they do not notice all of them (e.g., Suh & Trabasso, 1993). Van den Broek (1990) argued that the extent to which readers seek interconnections depends on their particular "standard of coherence" (van den Broek, Risden, & Husebye-Hartman, 1995; see also

[1]Within the constructionist model proposed by Graesser, Singer, and Trabasso (1994), readers do not necessarily attempt to make every possible connection. Instead, the number of connections established is determined by the goals of the reader.

van den Broek, Young, Tzeng, & Linderholm, 1996, for a thoughtful discussion of this issue in the context of the landscape model).

An alternative to these two contrasting extremes is the "memory-based text processing" view (e.g., Gerrig & McKoon, 1998; McKoon et al., 1996; McKoon & Ratcliff, 1995). According to this view, concepts currently in working memory make contact with other concepts processed earlier, as well as related concepts from general world knowledge through a fast-acting passive resonance process (cf. Gillund & Shiffrin, 1984; Hintzman, 1986; Kintsch, 1988; Murdock, 1983; Ratcliff, 1978; Ratcliff & McKoon, 1988). In this way, the reader gains ready access to relevant information from earlier portions of the text as well as from general world knowledge. This view may lead to a blending of the minimalist and constructionist views concerning what is generally available to the reader. With respect to the minimalist view, it further defines what was originally meant by "readily available" and provides a mechanism by which information not in the immediately preceding context can be readily available. With respect to the constructionist model proposed by Graesser et al. (1994), this passive resonance process is considered an uncontroversial bottom-up component of the comprehension process. However, central to the constructionist view is the assumption that passive activation mechanisms are not sufficient to account for all that is available to a reader; top-down processes that take into account readers' goals and strategies must also be considered. Thus, further specification of when information is likely to become available through a passive resonance mechanism will also serve to further specify when top-down processes must be considered.

In this chapter, we discuss some of the assumptions of a *resonance* model as it applies to text processing and present several recent findings consistent with the memory-based text-processing view (see Myers & O'Brien, 1998, for a more detailed presentation of the resonance model and its assumptions). We limit our discussion to factors governing the (re)activation of concepts during reading; and although we measure the impact this activation stage has on comprehension, we remain neutral with respect to how that information is evaluated and ultimately integrated into the evolving text representation. The model is based on the assumption that concepts derived from the sentence currently being processed, or concepts residing in working memory as a result of reading earlier portions of text, serve as signals to all of memory. The intensity of these signals may depend on the degree of attention given to concepts currently in focus, but the signal proceeds autonomously and is unrestricted. Concepts from earlier portions of the discourse representation as well as concepts from general world knowledge resonate as a function of the degree of match to the input. This match depends on the overlap of both *semantic* and *contextual* features among concepts. Memory elements that are contacted by the initial signal in turn signal to other memory elements. During this resonance process, activation builds and when the process stabilizes, the most active elements enter working memory. A critical aspect of the resonance process is that it is "dumb." Information that resonates sufficiently is returned to working memory independent of whether that information will ultimately facilitate or hinder further processing. In what follows, we consider some of the factors that influence the reactivation of concepts from earlier portions of the discourse, as well as the activation of related concepts from general world knowledge.

REACTIVATION OF CONCEPTS FROM EPISODIC MEMORY

Within the memory-based text-processing view and the resonance model, the primary factor that governs the reactivation of concepts from earlier portions of the discourse is the degree of featural overlap.[2] Although many factors (e.g., elaboration, strength of association in the episodic memory trace, referential distance) can interact with the degree of featural overlap to either increase or decrease the probability that any particular concept will be reactivated, a critical assumption of the resonance model is that without featural overlap reactivation will not occur. The role of featural overlap in the resonance process would explain Garrod and Sanford's (1982) finding that pronouns cannot typically be used to refer to backgrounded antecedents. Because pronouns contain little semantic content, the signal emanating from a pronoun is generally too limited to reactivate backgrounded antecedents.

Much of the research concerning the reactivation of antecedent information in response to an anaphoric reference is consistent with the memory-based text-processing view and the resonance model. For example, degree of featural overlap plays an important role in the reactivation of antecedents. McKoon and Ratcliff (1980; see also Dell, McKoon, & Ratcliff, 1983)

[2]There is no agreed upon definition of a feature, but we assume that two instances of the same concept within a text constitutes maximum overlap, and that more typical members of a category overlap more with the categorical noun than do less typical members. We also suspect that contextual features (e.g., Mensink & Raaijmakers, 1988) and features of the situation model constructed by the reader also play a role in the resonance process.

found that the time to reactivate an anaphor (e.g., *burglar*) was faster when the anaphor was a direct repetition of the antecedent than when it was a synonym (e.g., *criminal*). Presumably, repeating the noun provided greater featural overlap than the synonym. Similarly, Duffy and Rayner (1990; see also Garrod & Sanford, 1981) found that the time to resolve category name anaphors (e.g., *bird*) was longer when the antecedent was a low typical member of the category (e.g., *goose*) than when it was a high typical member (e.g., *robin*). Again low typical members share fewer features in common with the category than do high typical members (e.g., Rosch, Mervis, Gray, Johnsen, & Boyes-Braem, 1976).

A factor that is correlated with the probability of reactivation of antecedents is importance (O'Brien, 1987). The basis for this appears to be elaboration (e.g., Albrecht & O'Brien, 1991; Bloom, Fletcher, van den Broek, Reitz, & Shapiro, 1990; O'Brien & Myers, 1987; Trabasso & Sperry, 1985; Trabasso & van den Broek, 1985; van den Broek, 1990); concepts that are important components of a discourse tend to be mentioned several times within the text. When there are several related references to a concept, there is a buildup of activation on the concept (Anderson, 1983; Myers, O'Brien, Balota, & Toyofuku, 1984), increasing the probability and speed with which a signal from information currently in working memory will reactivate those concepts. Consistent with this view, O'Brien and Myers found that elaboration was the primary predictor of antecedent retrieval time. In a series of studies, O'Brien and his colleagues (e.g., O'Brien, 1987; O'Brien, Albrecht, Hakala, & Rizzella, 1995; O'Brien, Plewes, & Albrecht, 1990) found that although more recently occurring antecedent information is typically reactivated more quickly than more distant antecedent information, elaboration can override this "distance" effect. Consider the example passage in Table 4.1. O'Brien et al. (1990, 1995) found that elaborated antecedent information (e.g., *train*) was more quickly retrieved than was unelaborated information (e.g., *plane*), even when the elaborated antecedent was more physically distant than the unelaborated antecedent. Also, Albrecht and Myers (1998) found that simply adding an adjective modifier to elaborate a backgrounded episode increased the speed with which the information in that episode was subsequently reactivated.

An important aspect of the resonance process is that the signal is unrestricted; any concept that has features in common with the information in working memory should resonate in response to the signal emanating from that information. For example, when a text contains more than one potential antecedent to an anaphoric reference, it is likely that all candidates will be activated even though their reactivation will ultimately hinder the

TABLE 4.1

Example Passage From O'Brien et al. (1990, 1995)

Mark had grown up in the city but he had always wanted to live in the country. The first chance he got, he bought some land and moved there. It made him very happy not having to live in the crowded and noisy city. On holidays, he would travel by *train* into the city to visit his parents. While riding in it he liked to watch the countryside as it raced past him. Sometimes, the clackety clack it made on the tracks would put him to sleep. He'd wake up quickly though when they came to a crossing and it sounded the horn. Mark couldn't understand why people like his parents preferred to live in the city. Mark really enjoyed living in the country. He loved the open spaces and the clean fresh air. His brother had also moved out of the city and was now living in Colorado. Last summer Mark traveled by *plane* to visit him. He had loved looking down from it at the countryside and the clouds. Ever since Mark had moved to the country he had made a lot of friends. On Saturdays, he played golf with his neighbor. On the weekends, their families would get together for cookouts. One weekend they'd eat at Mark's and the next they would eat at his neighbor's. One night while they were talking, Mark's neighbor asked him how he traveled to his parents' (reinstates distant antecedent)/ his brother's (reinstates recent antecedent).

comprehension process. Consistent with this view, the time to read a categorical anaphor (e.g., *frozen vegetables*) is increased when the preceding text mentions more than one member of the category (e.g., *frozen peas and fresh corn*; Corbett, 1984; Mason, 1997). Similarly, O'Brien et al. (1990, 1995) found that reinstatement of a distant antecedent also resulted in the reactivation of a potential antecedent that had features in common with both the anaphor and the correct antecedent.

The ability of the repetition of a concept to reactivate distant information is nicely illustrated in several recent experiments. Greene, Gerrig, McKoon, and Ratcliff (1994; see also Gerrig & McKoon, 1998; McKoon et al., 1996) showed that the reunion of two characters in a discourse served to reactivate concepts that the characters had discussed in an earlier encounter. Consider the example in Table 4.2. In the Outsider Absent version of the story, Gloria's use of *she* in the pronoun sentence is unheralded; its antecedent (*cousin*) appeared several sentences earlier and is not readily available. However, Greene et al. found that Jane's return to Gloria in the reunion sentence made concepts they both knew about (e.g., *cousin*) more accessible than before the reunion sentence. Lea, Mason, Albrecht, Birch, and Myers (1998) extended this finding by demonstrating that it is not necessary for the characters to have shared knowledge of the target concept; all that is required is that one character associated with the concept be reinstated into the text. In the example presented in Table 4.3 Gloria does not have knowledge of the cousin. Nevertheless, Lea et al. still found that cousin was more accessible after the reunion sentence than before the reunion sentence. These findings are consistent with a passive resonance process in which the reinstatement of a character (or characters)

TABLE 4.2

Example Passage From Greene et al. (1994)

Introduction	
Jane was dreading her dinner with her cousin, Marilyn. She complained loudly to her roommate Gloria. "Every time I go to dinner at my cousin's I get sick." Gloria asked, "Why did you agree to go?" Jane said, "Because I'm too wimpy to say no." Jane went off to have dinner.	
Outsider Present Continuation	
When she arrived, Marilyn was just finishing the cooking. "You're in luck,," she said, "we're having fried squid." Jane knew she was in for a wonderful evening. The two of them sat down to dinner. After dinner, they talked for a while and then Jane left.	
Outsider Absent Continuation	
Gloria decided to cook something nice for herself for dinner. "As long as I'm home alone,," she thought, "I'll eat well." Gloria searched her refrigerator for ingredients. She found enough eggs to make a quiche. After dinner, she put the dishes in the dishwasher.	
Conclusion	
Reunion Sentence:	Gloria was still up when Jane arrived home about midnight.
Pronoun Sentence:	Gloria asked Jane, "Did she play you old disco records?"
Final Sentence:	Jane chuckled and said, "I can't get Disco Inferno out of my mind."
Recognition test word:	COUSIN

TABLE 4.3

Example Passage From Lea et al. (1998)

Introduction	
Jane and her roommate Gloria were leaving work. "Are you headed home?" asked Gloria. "No, but I'll see you later tonight,," replied Jane. Jane drove off to have dinner with her cousin, Marilyn. As she drove, she started to have regrets. She usually got sick when she ate at her cousin's. Jane wondered why she had agreed to go. She decided she was just too wimpy to say no.	
Continuation	
Meanwhile, Gloria went home and decided to cook something nice for herself. "As long as I'm home alone,," she thought, "I'll eat well." Gloria searched her refrigerator for ingredients. She found enough eggs to make a quiche. After dinner, she put the dishes in the dishwasher.	
Conclusion	
Reunion Sentence:	Gloria was still up when Jane arrived home about midnight.
Pronoun Sentence:	Gloria asked Jane, "Did she play you old disco records?"
Final Sentence:	Jane chuckled and said, "I can't get Disco Inferno out of my mind."
Recognition test word:	COUSIN

provides a signal to elements in an earlier episode and reactivates discourse entities associated with that episode.

A critical aspect of the resonance process is that the reactivation of information from inactive portions of the discourse representation is passive. Important evidence of the passive aspect would be a demonstration that comprehension is sensitive to distant information even when the text is locally coherent and searches of long-term memory are therefore not necessary. The results of several series of experiments (Albrecht & Myers, 1995, 1998; Albrecht & O'Brien, 1993; Cook et al., 1998; Hakala & O'Brien, 1995; Huitema, Dopkins, Klin, & Myers, 1993; Myers et al., 1994; O'Brien & Albrecht, 1992; O'Brien et al., 1998) have provided evidence that the resonance process activated relevant backgrounded information even when local coherence had been maintained. Consider the example passage in Table 4.4. We presented participants with passages containing a description of a particular characteristic of a protagonist (e.g., "Mary was a strict vegetarian"). Following several sentences that served to background

TABLE 4.4

Example of a Passage With Elaborated Characteristic of the Protagonist

Introduction
Today, Mary was meeting a friend Joan for lunch. She arrived early at the restaurant and decided to get a table. After she sat down, she started looking at the menu.
Consistent Elaboration
This was Mary's favorite restaurant because it had fantastic junk food. Mary enjoyed eating anything that was quick and easy to fix. In fact, she ate at McDonalds at least three times a week. Mary never worried about her diet and saw no reason to eat nutritious foods.
Inconsistent Elaboration
This was Mary's favorite restaurant because it had fantastic health food. Mary, a health nut, had been a strict vegetarian for 10 years. Her favorite food was cauliflower. Mary was so serious about her diet that she refused to eat anything that was fried or cooked in grease.
Neutral Elaboration
This was Mary's favorite restaurant because it had a nice quiet atmosphere. Mary frequently ate at the restaurant and had recommended it to all of her friends. She especially liked the cute tables and the country style cloths on them. It made her feel right at home.
O'Brien et al. (1998) Qualified Elaboration
Mary remembered that at a recent party, Joan played a joke by telling people that Mary had been a strict vegetarian for 10 years. Joan told everyone that Mary's favorite restaurant had fantastic health food. She said that Mary was a health nut and wouldn't eat anything that was fried or cooked in grease. She also claimed that Mary's favorite food was cauliflower.
Cook et al. (1998) Second Character Elaboration
This was Joan's favorite restaurant because it had fantastic health food. Joan, a health nut, had been a strict vegetarian for 10 years. Her favorite food was cauliflower. Joan was so serious about her diet that she refused to eat anything that was fried or cooked in grease.
Filler
After about 10 minutes, Mary's friend Joan arrived. It had been a few months since they had seen each other. Because of this Mary and Joan had a lot to talk about and chatted for over a half hour. Finally, they signaled the waiter to come take their orders. They checked the menu one more time. Mary and Joan had a hard time deciding what to have for lunch.
Critical Sentences
Mary ordered a cheeseburger and fries. She handed the menu back to the waiter.
Closing
Her friend didn't have as much trouble deciding what she wanted. She ordered and they began to chat again. They didn't realize there was so much for them to catch up on.

the characteristic, participants were presented with a target sentence in which the protagonist engaged in an action that was locally coherent but was either consistent, inconsistent, or neutral with respect to the earlier described characteristic (e.g., "Mary ordered a cheeseburger and fries"). Although the elaborated characteristic was not active in memory immediately prior to the target sentence (Myers et al., 1994), reading times on those sentences were significantly longer in the inconsistent condition than in either the consistent or neutral conditions. Consistent with the assumptions of the resonance model, when the reader encoded the sentence indicating that Mary ordered a cheeseburger and fries, that information sent a signal to all of memory. Memory traces relevant to Mary's eating habits resonated in response to that signal. When information indicating that Mary was a vegetarian was reactivated by this process, a global coherence break occurred. This in turn, resulted in a slowdown in reading as the reader attempted to reestablish coherence.

In an even stronger test of the memory-based text-processing view, O'Brien et al. (1998) included a "qualified" elaboration condition. An example is presented in Table 4.4. The elaboration in the qualification condition was rewritten to make clear to the reader that the critical characteristic was not true. If readers maintain an updated model of the protagonist, or if they actively search earlier portions of the discourse for relevant information, comprehension of the target sentence should not be disrupted; there is nothing in the overall content of the elaborated characteristic that is inconsistent with the target sentence. In contrast, assuming a passive resonance process, disconfirmed or false information should continue to resonate, become reactivated, participate in the integration of the target sentence, and cause comprehension difficulty. Consistent with the prediction based on the resonance process, reading times on the target sentence were significantly slower in the qualified elaboration condition. This finding strongly supports the assumption that the resonance process is both passive and dumb. The content of the intended message contained in the elaboration had no impact on the resonance process.

Another critical assumption of the resonance process is that the signal is unrestricted; any concept that shares features with concepts contained in the signal will resonate in response to that signal. This assumption is consistent with Garrod and Sanford's scenario-based model (1988, 1990; Sanford & Garrod, 1981). According to this model, the memory representation of a text can be partitioned into three distinct components: explicit focus, implicit focus, and long-term memory. Explicit focus contains explicitly mentioned discourse entities such as the protagonist. Implicit focus contains information from long-term memory that is discourse relevant; this would include characteristics of the protagonist, spatial-relational information, and relevant information from general world knowledge. Glenberg and Langston (1992) proposed that information in explicit and implicit focus may be connected through discourse pointers. The primary difference between the use of discourse pointers and the process assumed in memory-based models such as the resonance model (Myers & O'Brien, 1998; also see Goldman & Varma, 1995, and van den Broek et al., 1996) is subtle but important. Within the memory-based models, access to inactive portions of the discourse representation is gained strictly as a function of featural overlap. The ultimate relevance of that information does not play a role in the initial activation. In contrast, discourse pointers should direct the reader only to information in implicit focus that is relevant to the current contents of working memory. In order to contrast these two views, Cook et al. (1998) rewrote the elaboration section of the "Mary" passage so that the described characteristic was about a secondary character (e.g., Joan is described as a vegetarian: Table 4.4 contains an example). In a probe study, we found that when the target sentence described the primary character engaging in some action (e.g., Mary ordered a cheeseburger and fries), earlier described characteristics were reactivated even when they were associated with the secondary character. This finding is consistent with an unrestricted reactivation process; characteristics of the secondary character are reactivated because they share features in common with the action taken by the primary character. It is difficult to envision discourse pointers connecting the discourse entry for the primary character to characteristics of a secondary character.

When examining factors governing the accessibility of causal or goal-based information, a similar pattern emerges; initial reactivation appears to be independent of whether it is necessary or relevant to the integration process. For example, Table 4.5 contains a sample passage from Rizzella and O'Brien (1996). The last sentence ("He knew that once his father came home he would be in trouble.") provided a causal consequence for which there were two potential causal antecedents. One causal antecedent (e.g., Billy broke a small window) appeared in the immediately preceding context; a second, more distant causal antecedent appeared earlier in the passage (e.g., Billy lost his keys). Rizzella and O'Brien found that distant causal antecedents that were highly elaborated were reactivated when readers encountered a causal consequence even when the immediately preceding context provided the reader with a sufficient causal antecedent.

TABLE 4.5

Example Passage From Rizzella and O'Brien (1996)

Billy was walking home from school after playing a game of basketball. Billy looked for his keys to unlock the front door of his house. He searched everywhere but couldn't find the keys. He realized there was a big hole in his pocket. Now, he had no idea where to look. Billy shuddered when he recalled the warning his father gave him about being more responsible. His father told him that if he was not more responsible, he would ground Billy for an entire month. Billy needed to find another way to unlock the door. Billy broke a small window. The window fell in pieces on the ground. He knew that once his father came home he would be in trouble.

TABLE 4.6

Example Passage From Albrecht and Myers (1995, 1998)

Introduction
Mary was an advertising agent for a large firm in New York City. This evening, she was sitting at home working on a promotional piece for a blue jean company. This was her last project before she went on vacation. She needed to make an airline reservation tonight by midnight. If she didn't make the reservation tonight, she wouldn't get the special rate and might not even get a seat on the flight. She sat down on her (*leather*) *sofa* and looked through the telephone book.
Unsatisfied Goal
However, before she could confirm her reservation, she received a call from her boss. Apparently, the blue jean company wanted the ad done by tomorrow morning. Mary had to finish the project tonight and would have to confirm her reservation later.
Satisfied Goal
Just after she confirmed her reservation, she received a call from her boss. Apparently, the blue jean company wanted the ad done by tomorrow morning. Mary had to finish the project tonight and so she put away her reservation information.
Filler
She needed to decide on the colors for the layout. She set a black-and-white line drawing of the ad on her desk. Then she pulled out several color swatches. At first, she tried several different shades of green and red. However, she knew that the company would not like these colors. After much consideration, she selected royal blue and a light yellow. Now, she was done with the ad.
Context-Reinstatement Sentence
Exhausted, Mary sat down on the (*leather*) *sofa* for a moment.
Target Sentences
She was tired and decided to go to bed. She put on pajamas and washed her face.
Conclusion
She was certain she'd have no problem falling asleep.

This occurred even though reactivation of the distant cause was unnecessary to maintain causal coherence.

In an investigation of whether readers track goal-based information, Albrecht and Myers (1995) found that reactivation of an unsatisfied goal of a protagonist occurred only when the text currently in focus mentioned an object associated with the backgrounded goal. Table 4.6 contains a sample passage. Each passage began with the introduction of a goal episode (e.g., making airline reservations) along with specific contextual cues associated with the goal (e.g., *leather sofa*). The goal was either satisfied or unsatisfied and then backgrounded by a second, unrelated episode. If, following a shift in topic, *leather sofa* was again mentioned, the unsatisfied goal was reactivated. However, in the absence of this reinstatement of context surrounding the unsatisfied goal, there was no evidence that the goal was reactivated. In a subsequent study, Albrecht and Myers (1998) found that reinstating only *sofa,* without the modifier *leather* was also sufficient to reactivate the unsatisfied goal, but the reactivation process was delayed. If readers continually monitored goal-based information, then the reinstatement of the earlier context should not have been necessary to reactivate the unsatisfied goal; furthermore, increasing or decreasing the amount of information in the cue should not have influenced the time course of activation. Consistent with the memory-based view, the contextual cue served to reactivate the unsatisfied goal, and the more elaborate the cue (i.e., *leather sofa* vs. *sofa*), the faster the buildup of activation.

ACTIVATION OF CONCEPTS FROM GENERAL WORLD KNOWLEDGE

Although it is difficult to specify the reader's general world knowledge, we assume that it is activated in parallel with information in the discourse representation, and that resonance based on these two sources is processed simultaneously. Studies of memory of word lists (e.g., Perlmutter, Harsip, & Myers, 1976) and discourse (e.g., O'Brien & Albrecht, 1991) provide some support for our position that episodic information and general world knowledge both resonate in response to any signal emanating from working memory and interact to affect how discourse is comprehended. For example, in a study of antecedent retrieval, O'Brien and Albrecht showed that concepts absent from a passage, but having a strong semantic association to both the theme of the passage and an anaphoric phrase, were activated by the anaphoric phrase, and on some occasion they were incorrectly instantiated in response to an anaphoric phrase.

Albrecht and O'Brien (1991) found that concepts that were central to the theme of a discourse were recognized more quickly than concepts that were less central (see also O'Brien & Myers, 1987). However, O'Brien, Albrecht, Rizzella, and Hakala (1999) have shown that this was true only when the preexisting association of those concepts to the generic theme of the passage in general world knowledge was low. When the preexisting association of concepts to the generic theme of a passage was high, as would be the case for script-based narratives, the opposite pattern was obtained. The high degree of preexisting associations resulted in slower recognition times of the central concepts. Presumably, the recognition probe sends a signal to all of memory. As the amount of activation from general world knowledge increases, the ability to discriminate that activation from the activation obtained through the episodic memory trace decreases. Because central concepts would receive proportionally more activation from general world knowledge than less central concepts, recognition that those concepts were present in the episodic memory representation becomes more difficult.

The studies just cited demonstrated that subjects access information in both the discourse model and world knowledge during explicit memory tests. In a study currently in progress, we have begun to examine whether this same parallel access takes place during reading. Consider a text that explicitly states that several people purchased sandwiches for a picnic. After several sentences in which there was a clear change of topic, the same characters arrived at the picnic grounds and the next two sentences are:

> They took the picnic supplies out of the car.
> The sandwiches were delicious.

If only the last two sentences are read, readers construct a bridging inference; they understand that the sandwiches are a component of the picnic supplies (Haviland & Clark, 1974). Singer and Halldorson (1996) proposed that such inferences must be validated. Presumably, the speed of the validation process depends on the relation between the category (e.g., *picnic supplies*) and the exemplar in the next sentence (e.g., *sandwiches*). In several experiments we varied typicality; in our example, the exemplar in the second sentence would be *sandwiches* or *desserts*. In addition, we wrote passages in which the exemplar had been mentioned earlier. We varied the accessibility of that earlier established entity either by varying the distance between it and the second mention, or by varying the number of mentions (elaboration) when it was first introduced into the text.

Haviland and Clark (1974; also Kintsch & van Dijk, 1978) assumed that the reader attempts to match the given information to an antecedent in the memory representation of the previously processed discourse. *If no match is found,* the reader attempts to construct a bridge between the given information and information in the representation. This position would predict that only episodic accessibility (i.e., distance or elaboration) should have an effect on the time to read sentences like the second one in our example. On the other hand, van Dijk and Kintsch (1983) explicitly stated that the first sentence of such pairs immediately evokes world knowledge, a position consistent with Kintsch's (1988) assumption that propositions in the text spread activation to propositions in the knowledge base in an early construction stage. According to this view, the knowledge needed for the bridging inference is available before the matching antecedent. If the processor is satisfied by local coherence, we would expect an effect only of typicality. However, consistent with our position that information in both episodic and semantic memory resonate in response to what is currently being read, we have observed effects of both variables.

CONCLUDING REMARKS

The memory-based view of text processing and our instantiation of a resonance model (Myers & O'Brien, 1998) has a very strong bottom-up emphasis. It provides a mechanism that allows the reader ready access to information from both the episodic memory representation and general world knowledge; however, that access is not directed by the relevance of the information or by the goals of the reader. This view stands in contrast to the focus of constructionist models (e.g., Graesser et al., 1994; Singer et al., 1994) in which the emphasis has been on the reader's active role in searching for relevant information. Because the resonance process is passive, the reader never actively searches memory but evaluates and integrates information made available by the resonance process. This raises an important issue: What does the reader do when readily available information does not allow for comprehension to be complete? We assume that when the resonance process fails to provide sufficient information for comprehension to proceed smoothly, the reader refocuses on the current contents of working memory, sending a new signal to memory. This cycling process continues until sufficient information has been reactivated or until the reader decides to read further. Alternatively, the reader may choose to engage in active problem solving, a process beyond the scope of the resonance model but well captured within the constructionist model. The reader's motivation or "standards of coherence" (van den Broek et al., 1995) determine how much information will prove sufficient, and if or when to engage in some sort of active problem solving.

Currently, although the resonance process can account for many findings concerning the reactivation of information, there are aspects of the process that remain underspecified. For example, as with any model of memory based on featural overlap, the exact nature of a feature remains unclear. Also, an assumption of a strong version of the resonance process is that featural overlap is measured solely on the basis of preexisting semantic relations. However, this cannot be correct. For example, with the sample passage in Table 4.1, we would claim that "Mary ordered a cheeseburger and fries" activates that "Mary is a vegetarian." Clearly there is no direct featural overlap between *cheeseburger and fries* and *vegetarian.* There are two possible solutions. It could be that information about ordering a cheeseburger sends a signal to semantic memory, resulting in the activation of concepts such as "eating" and "meat"; these in turn activate vegetarian, which then makes contact with the episodic instantiation of that concept. Such a process would remove the need to assume the reader has developed a situation model in which a general profile of the protagonist is maintained; coherence would simply be checked against information activated in general world knowledge, which would then be mapped back onto the text base level of representation. However, we view this possibility as unlikely. It requires a considerable degree of mediated priming, and although there is evidence that such priming occurs, the strength and duration appears quite limited (e.g., Balota & Lorch, 1986; Lorch, Balota, & Stamm, 1986).

Alternatively, it may be that readers develop a situation model containing a rich representation of the protagonist as well as the situation described by the text. This representation would be constructed on the basis of explicitly stated information, general world knowledge, and inferences and elaborations generated by the reader. A well-established situation model would provide many of the semantic and contextual features necessary for reactivation of relevant information. Such a model may well have many of the characteristics of the causal structures and inferences envisaged by Trabasso and his colleagues (Trabasso, Secco, & van den Broek, 1984; Trabasso & Sperry, 1985; Trabasso & Suh, 1993; Trabasso & van den Broek, 1985; Trabasso, van den Broek, & Suh, 1989).

The data we have summarized makes a strong case for the sort of passive memory-based process proposed by McKoon et al. (1996), and instantiated in several recent simulation models, including our resonance model (Myers & O'Brien, 1998), Goldman and Varma's CAP-CI model (1995), and van den Broek et al.'s (1996) landscape model. Such models are as yet incomplete; they lack a full specification of mechanisms for evaluating the information accessed and integrating that information with the text representation. Some of those mechanisms are well specified within current constructionist models. Certainly any complete model of the comprehension process will contain many of the bottom-up components central to memory-based models as well as many of the top-down components central to constructionist models. Our goal is to further specify and understand those bottom-up processes that are likely to be an essential component of any complete model.

ACKNOWLEDGMENTS

The research presented in the chapter was supported by NSF Grants 96-30740 and 96-31040. The authors wish to express their thanks to Anne E. Cook for her helpful comments on an earlier draft of this chapter.

REFERENCES

Albrecht, J. E., & Myers, J. L. (1995). Role of context in accessing distant information during reading. *Journal of Experimental Psychology: Learning, Memory, and Cognition, 21,* 1459–1468.

Albrecht, J. E., & Myers, J. L. (1998). Accessing distant text information during reading: Effects of contextual cues. *Discourse Processes, 26,* 87–107.

Albrecht, J. E., & O'Brien, E. J. (1991). Effects of centrality on retrieval of text-based concepts. *Journal of Experimental Psychology: Learning, Memory, and Cognition, 17,* 932–939.

Albrecht, J. E., & O'Brien, E. J. (1993). Updating a mental model: Maintaining both local and global coherence. *Journal of Experimental Psychology: Learning, Memory, and Cognition, 19,* 1061–1070.

Anderson, J. R. (1983). *The architecture of cognition.* Cambridge, MA: Harvard University Press.

Balota, D. A., & Lorch, R. F. (1986). Depth of automatic spreading activation: Mediated priming effects in pronunciation but not in lexical decision. *Journal of Experimental Psychology: Learning, Memory, and Cognition, 12,* 336–345.

Bloom, C. P., Fletcher, C. R., van den Broek, P. W., Reitz, L., & Shapiro, B. P. (1990). An on-line assessment of causal reasoning during comprehension. *Memory & Cognition, 18,* 65–71.

Cook, A. E., Halleran, J. G., & O'Brien, E. J. (1998). What is readily available? A memory-based view of text processing. *Discourse Processes, 26,* 109–129.

Corbett, A. T. (1984). Pronominal adjectives and the disambiguation of anaphoric nouns. *Journal of Verbal Learning and Verbal Behavior, 23,* 683–695.

Dell, G. S., McKoon, G., & Ratcliff, R. (1983). The activation of antecedent information during the process of anaphoric reference in reading. *Journal of Verbal Learning and Verbal Behavior, 22,* 121–132.

Duffy, S. A., & Rayner, K. (1990). Eye movements and anaphor resolution: Effects of antecedent typicality and distance. *Language and Speech, 33,* 103–119.

Fletcher, C. R., & Bloom, C. (1988). Causal reasoning in the comprehension of simple narrative texts. *Journal of Memory and Language, 27,* 235–244.

Garrod, S. C., & Sanford, A. J. (1981). Bridging inferences in the extended domain of reference. In A. Baddley & J. Long (Eds.), *Attention and performance* (Vol. 9, pp. 331–346). Hillsdale, NJ: Lawrence Erlbaum Associates.

Garrod, S. C., & Sanford, A. J. (1982). The mental representation of discourse in a focussed memory system: Implications for the interpretation of anaphoric noun phrases. *Journal of Semantics, 1,* 21–41.

Garrod, S. C., & Sanford, A. J. (1988). Thematic subjecthood and cognitive restraints on discourse structure. *Journal of Pragmatics, 12,* 519–534.

Garrod, S. C., & Sanford, A. J. (1990). Referential processes during reading: Focusing on roles of individuals. In D. A. Balota, G. B. Flores d'Arcais, & K. Rayner (Eds.), *Comprehension processes in reading* (pp. 465–485). Hillsdale, NJ: Lawrence Erlbaum Associates.

Gerrig, R. J., & McKoon, G. (1998). The readiness is all: The functionality of memory-based text processing. *Discourse Processes, 26,* 67–86.

Gillund, G., & Shiffrin, R. M. (1984). A retrieval model for both recognition and recall. *Psychological Review, 91,* 1–67.

Glenberg, A. M., & Langston, W. E. (1992). Comprehension of illustrated text: Pictures help to build mental models. *Journal of Memory and Language, 31,* 129–151.

Goldman, S. R., & Varma, S. (1995). CAPping the construction-integration model of discourse comprehension. In C. A. Weaver, S. Mannes, & C. R. Fletcher (Eds.), *Discourse comprehension: Essays in honor of Walter Kintsch* (pp. 337–358). Hillsdale, NJ: Lawrence Erlbaum Associates.

Graesser, A. C., Singer, M., & Trabasso, T. (1994). Constructing inferences during narrative text comprehension. *Psychological Review, 101,* 371–395.

Greene, S. B., Gerrig, R., McKoon, G., & Ratcliff, R. (1994). Unheralded pronouns and the management of common ground. *Journal of Memory and Language, 33,* 511–526.

Hakala, C. M., & O'Brien, E. J. (1995). Strategies for resolving coherence breaks in reading. *Discourse Processes, 20,* 167–185.

Haviland, S. E., & Clark, H. H. (1974). What's new? Acquiring new information as a process in comprehension. *Journal of Verbal Learning and Verbal Behavior, 13,* 512–521.

Hintzman, D. L. (1986). "Schema abstraction" in a multiple-trace memory model. *Psychological Review, 93,* 411–428.

Huitema, J. S., Dopkins, S., Klin, C. M., & Myers, J. L. (1993). Connecting goals and actions during reading. *Journal of Experimental Psychology: Learning, Memory, and Cognition, 19,* 1053–1060.

Kintsch, W. (1988). The role of knowledge in discourse comprehension: A construction-integration model. *Psychological Review, 95,* 163–182.

Kintsch, W., & van Dijk, T. A. (1978). Towards a model of text comprehension and production. *Psychological Review, 85,* 363–394.

Lea, R. B., Mason, R. A., Albrecht, J. E., Birch, S. L., & Myers, J. L. (in press). Who knows what about whom: The role of common ground in accessing distant information. *Journal of Memory and Language, 39,* 70–84.

Lorch, R. F., Balota, D. A., & Stamm, E. G. (1986). Locus of inhibition effects in the priming of lexical decisions: Pre- or post-lexical access? *Memory & Cognition, 14,* 95–103.

Mason, R. A., (1997). *The role of mulitple antecedents in the time course of anaphor resolution.* Unpublished master's thesis, University of Massachusetts, Amherst.

McKoon, G., Gerrig, R., & Greene, S. B. (1996). Pronoun resolution without pronouns: Some consequences of memory-based text processing. *Journal of Experimental Psychology: Learning, Memory, and Cognition, 22,* 919–932.

McKoon, G., & Ratcliff, R. (1980). The comprehension processes and memory structures involved in anaphoric reference. *Journal of Verbal Learning and Verbal Behavior, 19,* 668–682.

McKoon, G., & Ratcliff, R. (1992). Inference during reading. *Psychological Review, 99,* 440–466.

McKoon, G., & Ratcliff, R. (1995). The minimalist hypothesis: Directions for research. In C. A. Weaver, S. Mannes, & C. R. Fletcher (Eds.), *Discourse comprehension: Essays in honor of Walter Kintsch* (pp. 97–116). Hillsdale, NJ: Lawrence Erlbaum Associates.

Mensink, G. J., & Raaijmakers, J. G. (1988). A model for interference and forgetting. *Psychological Review, 95,* 434–455.

Murdock, B. B. (1983). A distributed memory model for serial-order information. *Psychological Review, 90,* 316–338.

Myers, J. L., & O'Brien, E. J. (1998). Accessing the discourse representation during reading. *Discourse Processes, 26,* 131–157.

Myers, J. L., O'Brien, E. J., Albrecht, J. E., & Mason, R. A. (1994). Maintaining global coherence. *Journal of Experimental Psychology: Learning, Memory, and Cognition, 20,* 876–886.

Myers, J. L., O'Brien, E. J., Balota, D. A., & Toyofuku, M. (1984). Memory search without inference: The role of integration. *Cognitive Psychology, 16,* 217–242.

O'Brien, E. J. (1987). Antecedent search processes and the structure of text. *Journal of Experimental Psychology: Learning, Memory, and Cognition, 13,* 278–290.

O'Brien, E. J., & Albrecht, J. E. (1991). The role of context in accessing antecedents in text. *Journal of Experimental Psychology: Learning, Memory, and Cognition, 17,* 94–102.

O'Brien, E. J., & Albrecht, J. E. (1992). Comprehension strategies in the development of a mental model. *Journal of Experimental Psychology: Learning, Memory, and Cognition, 18,* 777–784.

O'Brien, E. J., Albrecht, J. E., Hakala, C. M., & Rizzella, M. L. (1995). Activation and suppression of antecedents during reinstatement. *Journal of Experimental Psychology: Learning, Memory, and Cognition, 21,* 626–634.

O'Brien, E. J., Albrecht, J. E., Rizzella, M. L., & Hakala, C. M. (1999). *Differing effects of centrality on retrieval of script-based versus narrative-based texts.* Manuscript in preparation.

O'Brien, E. J., & Myers, J. L. (1987). On the role of causal connections in the retrieval of text. *Memory & Cognition, 15,* 419–427.

O'Brien, E. J., Plewes, S., & Albrecht, J. E. (1990). Antecedent retrieval processes. *Journal of Experimental Psychology: Learning, Memory, and Cognition, 16,* 241–249.

O'Brien, E. J., Rizzella, M. L., Albrecht, J. E., & Halleran, J. G. (1998). Updating a mental model: A memory-based text processing view. *Journal of Experimental Psychology: Learning, Memory, and Cognition, 24,* 1200–1210.

Perlmutter, J., Harsip, J., & Myers, J. L. (1976). The role of semantic knowledge in retrieval from episodic long-term memories: Implications for a model of retrieval. *Memory & Cognition, 4,* 361–368.

Ratcliff, R. (1978). A theory of memory retrieval. *Psychological Review, 85,* 59–108.

Ratcliff, R., & McKoon, G. (1988). A retrieval theory of priming in memory. *Psychological Review, 95,* 285–408.

Rizzella, M. L., & O'Brien, E. J. (1996). Accessing global causes during reading. *Journal of Experimental Psychology: Learning, Memory, and Cognition, 22,* 1208–1218.

Rosch, E., Mervis, C. B., Gray, W. D., Johnsen, D. M., & Boyes-Braem, P. (1976). Basic objects in natural categories. *Cognitive Psychology, 9,* 382–440.

Sanford, A. J., & Garrod, S. C. (1981). *Understanding written language.* New York: Wiley.

Singer, M., Graesser, A. C., & Trabasso, T. (1994). Minimal or global inference during reading. *Journal of Memory and Language, 33,* 421–441.

Singer, M., & Halldorson, M. (1996). Constructing and validating motive bridging inferences. *Cognitive Psychology, 30,* 1–38

Suh, S., & Trabasso, T. (1993). Inferences during reading: Converging evidence from discourse analysis, talk-aloud protocols, and recognition priming. *Journal of Memory and Language, 32,* 279–301.

Trabasso, T., Secco, T., & van den Broek, P. W. (1984). Causal cohesion and story coherence. In H. Mandl, N. L. Stein, T. Trabasso (Eds.), *Learning and comprehension of text* (pp. 83–11). Hillsdale, NJ: Lawerence Erlbaum Associates.

Trabasso, T., & Sperry, L. L. (1985). Causal relatedness and importance of story events. *Journal of Memory and Language, 24,* 595–611.

Trabasso, T., & Suh, S. Y. (1993). Understanding text: Achieving explanatory coherence through on-line inferences and mental operations in working memory. *Discourse Processes, 16*(1–2), 3–34.

Trabasso, T., & van den Broek, P. (1985). Causal thinking and the representation of narrative events. *Journal of Memory and Language, 24,* 612–630.

Trabasso, T., van den Broek, P., & Suh, S. Y. (1989). Logical necessity and transitivity of causal relations in stories. *Discourse Processes, 12,* 1–26.

van den Broek, P. W. (1990). The causal inference maker: Towards a process model of inference generation in text comprehension. In D. A. Balota, G. B., Flores d'Arcais, & K. Rayner (Eds.), *Comprehension processes in reading* (pp. 423–446). Hillsdale, NJ: Lawrence Erlbaum Associates.

van den Broek, P. W., Risden, K., & Husebye-Hartmann, E. (1995). The role of readers' standards for coherence in the generation of inferences during reading. In R. F. Lorch, Jr. & E. J. O'Brien (Eds.), *Sources of coherence in reading* (pp. 353–373). Hillsdale, NJ: Lawrence Erlbaum Associates.

van den Broek, P. W., Young, M., Tzeng, Y., & Linderholm, T. (1996). The landscape model of reading inferences and the on-line construction of a memory representation. In H. Van Oostendorp & S. R. Goldman (Eds.), *The construction of mental representations during reading* (pp. 71–98). Mahwah, NJ: Lawrence Erlbaum Associates.

van Dijk, T. A., & Kintsch, W. (1983). *Strategies of discourse comprehension.* New York: Academic Press.

5

Revealing Inference Processes During Text Comprehension

Joseph P. Magliano
Northern Illinois University

Text understanding involves more than the construction and retrieval of a mental representation of the explicit text. Rather, understanding is achieved when readers are able to activate and apply their text-appropriate world knowledge in order to generate inferences regarding what the text is about. Furthermore, readers use their world knowledge in order to infer how elements of a text are related, and thus ensure that their mental representation of the text is coherent. Although there is consensus among discourse-processing researchers regarding the importance of world knowledge in understanding and achieving text coherence, there has been considerable theoretical conjecture and empirical research regarding what kinds of knowledge-based inferences readers normally generate during comprehension (Graesser, Singer, & Trabasso, 1994; Magliano & Graesser, 1991; McKoon & Ratcliff, 1992). Much of this debate has centered around (a) what inferences are necessary for achieving text coherence, (b) the extent to which readers generate inferences that are not necessary for achieving coherence, and more recently (c) what processes are involved in making inferences available to a reader. Although there has been much progress, the debates regarding these issues are far from resolved.

There has also been disagreement among researchers regarding the appropriate methodology for studying inference processes during reading (Keenan, Golding, Potts, Jennings, & Aman, 1990; Magliano & Graesser, 1991; Potts, Keenan, & Golding, 1988). This debate has centered around what methodologies most accurately expose those inference that occur "online," that is, while a person is actually reading a text. Lexical decision (e.g., Magliano, Baggett, Johnson, & Graesser, 1993; Potts et al., 1988), recognition priming (e.g., McKoon & Ratcliff, 1986; Potts et al., 1988; Suh & Trabasso, 1993; Trabasso & Suh, 1993), word naming (Potts et al, 1988), word stem completion (Whitney, & Williams-Whitney, 1990), reading time (e.g., Albrecht & O'Brein, 1993), question answering (Graesser & Clark, 1985), and thinking aloud (Trabasso & Magliano, 1996a, 1996b) methodologies have all been used to investigate online inference processes. Each of these methodologies has its pros and cons and circumstances when its use is either appropriate or inappropriate (see Magliano & Graesser, 1991, for a comprehensive review). Given that there is considerable debate regarding these methodologies, the most prudent approach for studying online inferences is to rely on converging methodologies (Keenan, Potts, Golding, & Jennings, 1990; Magliano & Graesser, 1991; Potts et al., 1988).

In keeping with the notion that it's important to utilize convergent methodologies when studying online inference processes, Magliano and Graesser (1991) argued for a three-pronged approach for studying inference generation. The three-pronged approach coordinates theories of discourse processing, verbal inference elicitation methods (e.g., thinking aloud and question answering), and online behavioral measures. The theories provide a principled foundation for making predictions regarding what kinds of inferences should be generated online. Verbal inference elicitation methods are used to empirically uncover potential inferences. Finally, online behavioral measures provide rigorous methods for assessing empirically whether particular classes of inference are generated online. Convergence among the three prongs strengthens one's conclusions about inference processes. For example, theories of discourse processing can be validated by both elicitation methods and online behavioral measures. There can also be convergence between elicited inferences and the online behavioral measures. For example, elicited inferences can be used to predict sentence reading times (Olson, Duffy, & Mack, 1984; Trabasso & Suh, 1993). Elicited inferences can also provide the materials for priming tasks, such as lexical decision tasks (Long & Golding, 1993; Magliano et al., 1993) and recognition tasks (Suh & Trabasso, 1993; Trabasso & Suh, 1993). This ensures that the materials for these tasks involve actual inferences that participants could generate during comprehension. Accordingly, the three-pronged approach is a unique, powerful, and comprehensive method for studying online inference processes.

Tom Trabasso has been an important pioneer and advocate of the three-pronged approach (Suh & Trabasso, 1993; Trabasso & Suh, 1993).

Trabasso's causal network model (Trabasso & Sperry, 1985; Trabasso, van den Broek, & Suh, 1989) has provided an important and widely used theoretical tool for analyzing the causal structures that are inherent in texts and for predicting the causal inferences readers should generate during comprehension. Trabasso has also been particularly instrumental in developing think-aloud methods as a way to elicit inferences and study online inference generation. Trabasso and Magliano (1996a) proposed a rigorous method for analyzing think-aloud protocols collected during comprehension. Their approach is unique because it allows (a) an assessment of the inferential content of the think-aloud protocols, (b) an assessment of the information source for a particular inference, and (c) an assessment of the memory operations that make that information available to a reader. As is discussed in detail later in this chapter, there is a great deal of convergence between the causal network model, think-aloud protocols, and independent, online behavioral measures (Magliano, Zwaan, & Graessser, 1998; Suh & Trabasso, 1993; Trabasso & Suh, 1993).

The purpose of this chapter is to discuss the three-pronged approach with an emphasis on Trabasso's important contributions. The first section provides a brief description of the causal network model. This model is discussed in some detail because it provided an important theoretical foundation for much of the research discussed in this chapter. As such, it provided the theory prong of the three-pronged approach that was adopted in conducting this research. The second section discusses a comparison between two important inference elicitation methods, namely question answering and thinking aloud. This comparison provides additional converging evidence with respect to the inferences that are generated online. Furthermore, it provides some indications regarding when the use of each inference elicitation method is appropriate. Finally, in the third section, a few studies are discussed that demonstrate convergence between the causal network model, think-aloud protocols, and independent, online behavioral measures. These studies were either conducted by Trabasso, conducted in collaboration with him, or inspired by his research.

THE CAUSAL NETWORK MODEL

The first prong of the three-pronged approach is theory. Theories of inference generation provide principled bases for determining what inferences are generated online, when those inferences should be generated, the mechanisms underling the generation of those inferences, and an account of how those inferences function to promote understanding. The *causal network* model proposed by Trabasso et al. (1989) provides a theoretical account of some of these aspects of inference generation. An important assumption of this model is that understanding is achieved primarily through causal reasoning because these causal inferences provide the glue that bind story elements together. As such, many of the inferences that are generated online are causal in nature. In order to identify what and when causal inferences should be generated when reading a story, the causal network model focuses on providing a basis for determining causal relationships between sentences in stories. According to the causal network model, story sentences can be classified according to how they fit into an episodic structure. Episodes consist of a set of categories of story units (cf. Stein & Glenn, 1979). These categories include settings (S), events (E), goals (G), attempts (A), outcomes (0), and reactions (R). Table 5.1 shows the episodic categories of each sentence in "Jimmy and the Bike," a typical story used in this research (Goldman & Varnhagen, 1986; Suh & Trabasso, 1993; Trabasso & Suh, 1993; van den Broek, 1989). The causal network model also specifies the types of causal relationships that can occur between these categories. Settings enable all other categories. Events can physically cause events and psychologically cause characters to have goals, and reactions. Goals can motivate other goals and attempts. Attempts can enable attempts and physically cause outcomes. Outcomes, as well as events, can

TABLE 5.1

The Hierarchical Version of "Jimmy and the Bike"

Story Sentence	*Episodic Category*
1. Once there was a boy named Jimmy.	Setting
2. One day, Jimmy saw his friend Tom riding a new bike.	Event
3. He wanted to buy a bike.	Goal
4. Jimmy spoke to his mother.	Attempt
5. Jimmy's mother refused to get a bike for him.	Outcome
6. He was very sad.	Reaction
7. Next day, Jimmy's mother told him that he should have his own savings.	Event
8. Jimmy wanted to earn some money.	Goal
9. He asked a job at a nearby grocery store.	Attempt
10. He made deliveries for the grocery store.	Attempt
11. He earned a lot of money.	Outcome
12. Jimmy went to the department store.	Attempt
13. He walked to the second floor.	Attempt
14. Jimmy bought a new bike.	Outcome

psychologically cause characters have reactions and goals. Outcomes can also enable attempts. The causal relationships specified by the model for a particular story are based on the logical criteria of necessity and weak sufficiency in the circumstances (Mackie, 1980). An Event A is necessary for an Event B if Event B will not occur in the story without Event A. For example, Jimmy would not have wanted to get a new bike had he not seen his friend riding a new one. Furthermore, if Event A occurs in the story, then Event B is likely to follow. For example, given that Jimmy saw his friend riding a new bike, it is likely he would want a new one in the context of the story.

Figure 5.1 shows a network representation of the causal relationships between sentences for the story in Table 5.1. The arrows depict causal relationships between story sentences. An arrow between sentences S(i-1) and S(i) indicates that S(i) is causally connected with its immediate context (i.e., there is causal continuity). S(i) is distally related with the prior story context when it is causally connected with S(i-n) where n is greater than 1. A majority of the distal connections in these kinds of stories involve the thematic goals introduced in the first and second episodes of the stories. However, there are distal causal connections between story sentences that do not involve the thematic goals.

The causal network analysis developed by Trabasso and his colleagues is quite robust in that it can predict a variety of both online and off-line memory behaviors (see van den Broek, 1995, for a comprehensive review). With respect to off-line measures of memory, the causal network model can predict the content of free recall protocols (e.g., Goldman & Varnhagen, 1986; Trabasso & van den Broek, 1985; van den Broek, Rohleder, & Narváez, 1996), story summary protocols (e.g., van den Broek et al., 1996) judgments of importance for story sentences (Trabasso & Sperry, 1985), and judgments of causal strength between story elements (Trabasso et al, 1989). As is discussed in detail in the third section of this chapter, the causal

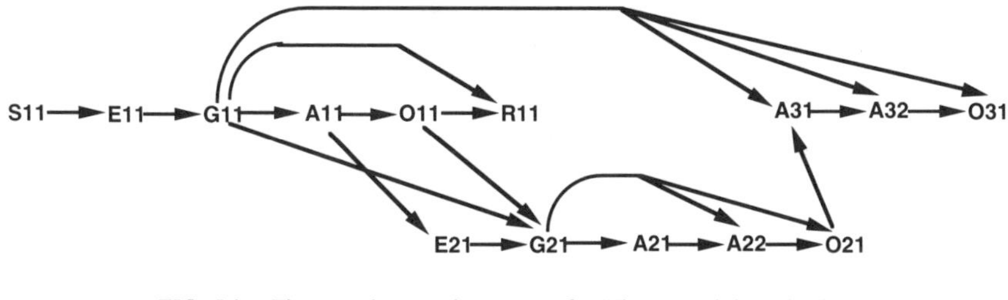

FIG. 5.1. The causal network structure for "Jimmy and the Bike."

network model can predict a variety of online behaviors, such as sentence reading times (Magliano et al., 1998; Trabasso & Suh, 1993), recognition priming latencies (Lutz & Radvansky, 1997; Suh & Trabasso, 1993; Trabasso & Suh, 1993; van den Broek & Lorch, 1993), lexical decision latencies (van den Broek et al., 1996), goodness of fit judgments for story sentences (Magliano et al., 1998), and the inferences produced during thinking aloud (Magliano et al., 1998).

The analysis outlined by the causal network model is both manageable to use and versatile. Not only can it be applied to experimenter-generated text, such as the one shown in Table 5.1, but it can also be applied to complex texts, such as folk tales (Trabasso & Sperry, 1985) and actual literary texts (van den Broek et al., 1996). It has also been used to analyze the causal structure of both think-aloud protocols produced when watching a movie (Magliano, Trabasso, & LeSage, 1995) and memories for real-life experiences (Stein, Folkman, Trabasso, & Richards, 1997). The model provides a basis for understanding how people make sense of narrative events depicted in text and film, and experienced in real life.

TWO MAJOR METHODS FOR REVEALING POTENTIAL INFERENCES: THINKING ALOUD AND QUESTION ANSWERING

The second prong of the three-pronged approach is verbal inference elicitation methods. Question-answering (Graesser, 1981; Graesser & Clark, 1985) and think-aloud methodologies (Collins, Brown, & Larkin, 1980; Coté & Goldman, 1998; Coté, Goldman, & Saul, 1998; Ericsson & Simon, 1994; Olson, Duffy, & Mack, 1984; Suh & Trabasso, 1993; Trabasso & Magliano, 1996a, 1996b; Trabasso & Suh, 1993; Zwaan & Brown, 1996) are two of the major techniques that have been used to expose (elicit) inferences that are potentially available during reading. There has never been a comparison between these two methods regarding the inferences that they expose during reading. Such a comparison may provide additional convergence in determining which inferences are generated online. This section contains a comparison of these two methodologies. Data are presented from a study that was conducted in collaboration with Trabasso and Graesser that assessed the extent to which question–answer and think-aloud protocols overlap with regards to the inferences that they expose. Before describing these studies, an overview of each methodology is provided.

The question-answering methodology is particularly useful for exposing a relatively complete set of world knowledge that is relevant to a specific story and potentially accessible to a reader during online and off-line processing. Graesser and his colleagues have been the biggest proponents of the question-answering methodology (Graesser, 1981; Graesser & Clark, 1985). In a question-answering methodology, readers read a text and generate answers to specific queries for each sentence or clause in the order in which the sentences appear in the text. In previous studies, Graesser has probed readers with three types of questions: *why, how,* and *what-happens-next* (WHN). In principle, however, it is possible to ask other types of questions. Responses to the sentences can be compared with responses generated when the sentences are queried outside the context of comprising a narrative. This approach exposes the larger set of world knowledge associated with the idea units that occur in a text independent from that text.

The question-answering methodology exposes a wide range of inferences. Answers include goals, intentional actions, involuntary events, states, and the styles in which actions and events occur (Graesser & Clark, 1985). Both the question type (e.g., why, how, and WHN questions) and type of information that is being queried (e.g., involuntary events and intentional actions) have an impact on the kinds of inferences that are exposed in question answering. Why and how questions regarding an involuntary event (e.g., The barn caught on fire) expose the causal antecedents of that event (e.g., The barn was struck by lightening). WHN questions regarding an involuntary event expose the causal consequences of that event (e.g., The barn burned to the ground). Why questions regarding an intentional action (e.g., Bob hit his best friend) expose the superordinate goal that motivated the action (e.g., to get revenge) and the events that may have precipitated the action (e.g., Bob caught his friend in bed with his wife). How questions regarding an intentional action expose the subordinate actions, the style and manner, and/or the tools involved in completing the queried action (e.g., Bob hit his friend very hard, Bob used his fist). WHN questions regarding intentional actions expose the outcomes of that action (e.g., Bob was arrested and charged with assault). Why questions expose the lion's share of the inferences that occur during online comprehension for both unintentional events and actions (Graesser & Clark, 1985; Graesser, Haberlandt, & Koizumi, 1987).

Think-aloud methodologies are becoming increasingly accepted as valid tools for studying comprehension processes (Ericsson & Simon, 1994; Long & Bourg, 1995; Pressley & Afflerbach, 1995; Trabasso & Magliano, 1996a, 1996b; Whitney & Budd, 1996; Zwaan & Brown, 1996). With a think-aloud methodology, readers are asked to report whatever thoughts come to mind as they read a text. In order to expose online inferences, readers are asked to report thoughts after reading text sentences or clauses. However, one could collect think-aloud protocols after participants read larger segments, such as paragraphs or chapters. Think-aloud protocols reveal what is available to working memory, accessible to consciousness, and codeable in language (Ericsson & Simon, 1994). Think-aloud protocols show how available information is used in an *effortful* search for meaning during comprehension (Bartlett, 1932; Coté, Goldman, & Saul, 1998; Graesser et al., 1994; Stein & Trabasso, 1985). As such, think-aloud protocols expose conscious, strategic processing.

Trabasso and Magliano (1996a) developed a think-aloud protocol analysis that is particularly useful for studying online inference processes. This analysis reveals the kinds of inferences that are potentially generated online, what information sources are the bases for those inferences, and the memory operations that are carried out to make the necessary information available for understanding. The Trabasso and Magliano analysis distinguishes between three broad categories of inferences. Explanations are backward oriented in the narrative timeline with respect to a focal sentence and provide reasons why a story event or action has occurred. Associations are concurrent in time with respect to the focal sentence and typically provide elaborative and descriptive detail (e.g., the age of a character, the size or color of an object, or an instrument used to accomplish an action). Predictions are forward oriented in narrative time with respect to the focal sentence and provide the consequences of narrative events and actions. Inferences can be based on world knowledge, the prior text, or previously generated knowledge-based inferences from the long-term memory representation for the text. The information necessary to generate an inference can be activated from world knowledge, reinstated from the long-term memory representation for a text, or maintained and carried over in working memory from the immediately prior sentence.

Trabasso and Magliano (1996a) found that explanations, associations, and predictions occurred at different frequencies during thinking aloud. Explanations, associations, and predictions accounted for approximately 70%, 19%, and 11% of the inferences, respectively. Thus, explanations predominated when examining the inferences produced while thinking aloud. With respect to how the various information sources were inferentially used, prior text information, world knowledge, and prior inferences were primarily activated as a basis for explanatory inferences. Associative inferences, however, tended to be based on world knowledge rather than prior text or previously generated inferences. Finally, readers retrieved information from the long-term memory representation for the text or maintained information in working memory over successive sentences primarily to generate explanatory inferences. Trabasso and Magliano's results are consistent with the assumption that understanding is achieved largely through explanatory reasoning (e.g., Graesser et al., 1994). Furthermore, the general pattern of results reported by Trabasso and Magliano has been replicated by other researchers who collected think-aloud protocols in the context of narrative (Zwaan & Brown, 1996), as well as expository (Coté & Goldman, 1998; Coté et al., 1998) texts.

Because both question answering and thinking aloud elicit inferences potentially available during comprehension, it is interesting to examine the extent to which the two methods elicit the same inferences. Trabasso, Graesser, and I addressed this issue by collecting think-aloud protocols on a set of short narratives for which Graesser and Clark (1985) had previously collected responses to why, how, or WHN questions for each sentence. The Trabasso and Magliano (1996a) analysis was applied to the protocols. In order to assess the extent to which the question-answering and the think-aloud protocols overlapped in their inferential content, we sampled 22 sentences across the four stories. The 22 sentences were chosen because they conveyed unambiguously either an intentional action or unintentional event. This distinction was important because Graesser and Clark showed that different kinds of causal inferences are elicited at actions and events, but this distinction is not discussed here. Twelve sentences depicted a character engaged in an intentional action and 10 sentences depicted an involuntary event. For each of these 22 sentences, an overlap score was calculated between the statements produced during thinking aloud and those produced when answering why, how, and WHN questions. Overlap scores calculate the ratio of the frequency of overlapping inferences from the think-aloud and question-answering protocols to the total frequency of statements for both sets of protocols (see Graesser, 1981, for a more detailed discussion of overlap scores). It was assumed that the overlapping inferences reflected those inferences that were likely to occur online.

The volume of inferences in statements that were produced during thinking aloud and question answering was about the same. Readers produced a mean number of 1.89, 2.21, 2.08, and 1.89 inferences per statement when thinking aloud, answering why questions, answering how questions, and when answering WHN questions, respectively. However, there were differences in the overlap between thinking aloud and answering why, how, and WHN questions. There was an overlap score of .21, .09, and .14 between the think-aloud protocols and why, how, WHN-question protocols, respectively. A one-way repeated-measures ANOVA (analysis of variance) indicated that these overlap scores were significantly different from one another ($F(2,42) = 9.07$, MSE = .01, $p < .01$). Planned comparisons using the ANOVA error term indicated that the overlap scores for why questions were higher than that of how questions ($t(42) = 3.98, p < .01$) and WHN questions ($t(42) = 2.32, p < .05$). The overlap scores between how and WHN questions were not significantly different from one another. As stated previously, why questions expose causal explanations. Given that the overlap scores were highest for thinking aloud and why questions, the overlap scores support the claim that understanding is based largely on explanatory reasoning.

Question answering and thinking aloud are useful for different purposes. Question answering is very useful when researchers want to expose a particular type of inference category (e.g., why questions expose explanatory inferences) or when researchers want to expose a relatively complete set of knowledge that is relevant to a story and potentially available during online and off-line processing (Graesser & Clark, 1985). In contrast to the question-answering methodology, with a think-aloud methodology, readers are not given specific instructions regarding what they should talk about. As such, thinking aloud may be less obtrusive than question answering. Thinking aloud may be particularly useful when one wants to assess the frequency at which different inferences occur when readers understand a text without specific inference strategies (Trabasso & Magliano, 1996a, 1996b). Furthermore, thinking aloud is an important tool when researchers are interested in studying a broad range of online processes and not just knowledge-based inferences, per se. The think-aloud methodology also provides an important tool for exposing the problem-solving, metacognitive, and evaluative strategies and processes that occur during online comprehension (Collins et al., 1980; Pressley & Afflerbach, 1995; Trabasso & Magliano, 1996a. 1996b).

CONVERGENCE BETWEEN THE THREE PRONGS: BRIDGING THE GAP BETWEEN POTENTIAL AND ONLINE INFERENCES

The third prong of the three-pronged approach is online behavioral measures, such as sentence reading times and priming latencies. In the previous two sections, I discussed Trabasso's contributions to two of the prongs of the three-pronged approach, namely theory and inference elicitation methods. In this section, research is discussed that was either conducted by Trabasso or inspired by his work and that demonstrates convergence between theory, verbal inference elicitation methods, and online behavioral measures. The strength of the three-pronged approach is that it relies on convergence between theory, inference elicitation methods, and online behavioral measures. Such convergence provides a more complete and convincing picture of text understanding than relying on convergence between theory and online behavioral measures alone (Magliano & Graesser, 1991). There are a growing number of discourse-processing researchers who have adopted the three-pronged approach to study online inference processes (Langston, Trabassso, & Magliano, in press; Long & Golding, 1993; Lutz & Radvansky, 1997; Magliano et al., 1993; Magliano et al., 1998; Millis & Graesser, 1994; Millis, Morgan, & Graesser, 1990; Suh & Trabasso, 1993; Trabasso & Suh, 1993). In this section, some of this research is reviewed with an emphasis on studies that are relevant to the causal network model.

Convergence Between the Causal Network Model and Inferences Produced While Thinking Aloud

In this section, a few studies are discussed that have demonstrated convergence between the causal network model and the thoughts produced while thinking aloud. Again, such convergence provides an understanding of when theory can predict behavior, namely the inferences produced during reading. Suh and Trabasso (1993) constructed stories that were derived by the causal network model (Trabasso et al., 1989). There was a sequential and hierarchical version of each story. Table 5.1 contains the hierarchical version of the story "Jimmy and the Bike." In the first episode of the hierarchical version, a goal was introduced but not obtained (e.g., Jimmy's mother refused to buy him a bike), whereas it was obtained in the sequential version (Jimmy's mother buys him a bike). In the second episode of both versions, a second goal was introduced and achieved (e.g., Jimmy wants to earn money). In the hierarchical version, the second goal could be interpreted as a being subordinate to the first goal (e.g., Jimmy wants to earn money because he wants to buy a bike), whereas it could be interpreted as a new, unrelated goal in the sequential version. The third episode depicted two to three attempts that were followed by a successful outcome. For the hierarchical version, the outcome in Episode 3 involved achievement of the first goal, whereas it involved some other goal in the sequential version. Suh and Trabasso assessed the extent to which the causal network model predicts when readers mentioned the first and second goals when thinking aloud. For example, as predicted by the causal network model, there was a significantly higher probability of mentioning the first goal when thinking aloud about the second goal statement for the hierarchical versions of the stories than the sequential versions. There was also a higher probability of mentioning the first goal when thinking aloud about the attempts depicted in the third episode for the hierarchical versions than the sequential versions. The results of Suh and Trabasso clearly demonstrated that readers were highly sensitive to the causal connections that were identified through a causal network analysis when thinking aloud. The results of Suh and Trabasso have been replicated in other laboratories (e.g., Lutz & Radvansky, 1997).

In another set of analyses that demonstrated the convergence between the causal network analysis and generated inferences, I assessed the extent to which the causal connections identified by a causal network analysis are correlated with a broader range of inferences produced while thinking aloud. The theory would predict that when the story contains causal explanations for events that occur, readers should connect these statements; when causes are implicit or absent readers should generate new knowledge-based inferences (Trabasso et al., 1989; van den Broek, 1990). This analysis involved the think-aloud protocols for the study that was conducted in collaboration with Trabasso and Graesser that was discussed earlier. Again, participants read and thought aloud about story sentences, one at a time, at their own pace. The stories included were those that were investigated by Graesser and Clark (1985) and four additional stories selected from Suh and Trabasso (1993). The causal structure inherent in these stories was identified using the causal network analysis. The number of causal antecedents and causal consequences was identified for each story sentence ($N = 137$). The think-aloud protocol analysis developed by Trabasso and Magliano (1996a) was also adopted to analyze the inferential content of the protocols. As consistent with van den Broek (1990), a distinction was made between text-connecting and knowledge-based inferences. The number of new explanations, predictions, and associations that occurred at each sentence was identified. An inference was considered a new knowledge-based inference if it was not generated in any of the prior sentences by any participant and did not contain information from the focal sentence or any sentence in the prior context. There were 4.27 new explanations, 1.75 new predictions, and 2.23 new associations generated for each sentence. Thus, as is consistent with Trabasso and Magliano (1996a, 1996b), explanations predominate new knowledge-based inferences. The number of text-connecting inferences that were produced during thinking aloud was also calculated for each story sentence. Text-connecting inferences contained information that was mentioned in a prior sentence. Text-connecting inferences were almost exclusively explanations and were assumed to occur when readers made a causal connection between the focal sentence and some prior sentence. There were .86 text-connecting inferences generated per sentence.

A series of multiple regression analyses were conducted on the number of new knowledge-based explanations, associations, predictions, and text-connecting inferences generated at each sentence. The predictor variables in these analyses were the number of causal antecedents and the number of causal consequences for each sentence. The beta-weights and variance explained by each equation are shown Table 5.2. The analyses on text-connecting explanations accounted for a significant 28% of the variance. This analysis revealed that the occurrence of text-connecting explanations during thinking aloud increased when there was a causal antecedent(s) identified by the causal network model. The analyses for knowledge-based explanations and associations each revealed a similar pattern and both accounted for a significant 4% of the variance. Knowledge-based explanations and associations tended *not* to occur when there was an antecedent(s) identified by the causal network model. The analysis for knowledge-based predictions was not significant. These data suggest that the causal network model can be used to make predictions regarding when a broad range of inference categories are generated while reading a text. Specifically, readers relied on the prior context to explain a focal event when it was causally connected to the prior context, whereas they relied on world knowledge when it was not readily causally integrated into the prior context.

TABLE 5.2

Regression Analyses Assessing the Relationship Between Causal Structure and the Occurrence of Text-Based Explanations, Knowledge-Based Explanations, Associations, and Predictions

	INFERENCE TYPE			
Predictor Variables	*TB Explanations*	*KB Explanations*	*KB Associations*	*KB Predictions*
Antecedents	.52*	–.19*	–.19*	–.11
Consequences	.04	–.04	–.05	.09
VARIANCE EXPLAINED	28%*	4%*	4%*	NS

Note. * = significant at .05 alpha, TB = text connecting, and KB = knowledge based.

Convergence Between Theory, Inferences Elicitation Methodologies and Sentence Reading Times

One three-pronged approach involves using verbal protocols to predict sentence reading times in order to assess what kinds of inferences readers generate online. This approach involves first deriving a set of theoretically interesting variables from the verbal protocols that are collected on a series passages. A separate group of participants read these same passages silently and sentence reading times are collected. The theoretically interesting predictor variables are then incorporated in a regression analysis in order to predict sentence reading times. This regression analysis yields a directional beta-weight for each predictor variable (see Graesser & Riha, 1984, for a more complete discussion of the using of a regression analysis to predict sentence reading times). These beta-weights indicate whether or not there is a significant relationship between a variable and reading times, as well as the direction of that relationship. With such an approach, Graesser and his colleagues have demonstrated that the explanations produced when answering why questions predict sentence reading times in the context of understanding both narrative (Graesser et al., 1987) and expository texts (Millis et al., 1990).

In line with such an approach, I conducted a series of regression analyses in order to further assess the similarities and differences between question answering and thinking aloud in revealing online inferences. The purpose of this analysis was to assess the extent to which thinking aloud and question answering converge to predict reading times. Convergence between these analyses should further reveal which inferences occur online. Given the results of the overlap analysis between thinking aloud and questions answering, it was expected that the convergence in these analyses would primarily occur for explanatory inferences. A set of predictor variables were derived from the Graesser and Clark (1985) question-answering protocols and from the think-aloud protocols that were collected on these same passages. One set of predictor variables involved identifying the number of new knowledge-based inferences that were produced at each sentence when answering questions and when thinking aloud. With respect to the question-answering protocols, this involved the number of new inferences produced when answering why, how, and WHN questions for each sentence. With respect to the think-aloud protocols, this involved identifying the number of new knowledge-based explanations, associations, and predictions that were produced at each story sentence. A second set of variables involved text-connecting inferences. The first of these variables involved the number of text-based explanations that were produced when answering a why question and produced while thinking aloud for each sentence. The second of these variables involved identifying those sentences that substantiated a prior prediction that was generated by any participant when answering a WHN question and when thinking aloud. One set of regression analyses were conducted using these variables as derived from the question-answering protocols and the other set involved the think-aloud protocols.

Table 5.3 shows the beta-weights and variance explained by each predictor variable from these analyses. There is a great deal of convergence between the analysis based on the think-aloud protocols and that based on the question-answering protocols. Sentence reading times decreased when a sentence was readily integrated with the prior story context, that is, when it could be explained by the prior context or when it matched a prior prediction. Sentence reading times, on the other hand, increased when readers had to rely on world knowledge in order to explain the focal sentence.

These analyses are complementary to regression analyses reported in the last section: The text-connecting explanations had a higher likelihood of occurring when a focal sentence was causally connected with the prior context (as identified through a causal network analysis), whereas knowledge-based explanations and associations had a higher likelihood of occurring when the focal sentences were *not* causally connected with the prior context. Both the regression analyses reported previously and the regression analyses reported here converge to paint a consistent picture of how a focal sentence is integrated with prior text. Sentence processing is facilitated to the extent that a sentence is readily causally integrated into the prior context.

TABLE 5.3

Beta-Weights and Variance Explained for the Theoretically Important Predictor Variables

	Thinking Aloud		*Question Answering*	
Inference Category	*Beta-Weights*	*Variance Explained*	*Beta-Weights*	*Variance Explained*
New Inferences				
Explanations	.18	3%	.19	3%
Associations	.37	12%	NS	
Predictions	NS		NS	
Text Connecting				
Explanations	–.23	4%	–.19	3%
Prior Expectation	–.17	3%	–.26	6%

Note. NS = not significant.

It is when readers cannot readily integrate a sentence into the prior context that they must rely heavily on the relevant world knowledge to construct meaning. The interpretations of a focal sentence based primarily on knowledge-based inferences may require relatively more effort to integrate into the existing memory structure, which results in longer sentence reading times. Although these analyses are purely correlational, the convergence is consistent with the view that causal-explanatory reasoning provides the foundation of understanding (Graesser et al., 1994; Trabasso et al., 1989). This reasoning provides a fundamental basis for understanding how a focal sentence fits into the prior text context.

Convergence Between Theory, Inference Elicitation Methods, and Online Priming Methodologies

A final convergence discussed here involves priming paradigms, such as lexical decision (e.g., Long & Golding, 1993; Magliano et al., 1993) or recognition priming (e.g., Suh & Trabasso, 1993; Trabasso & Suh, 1993). This approach is used primarily to assess the extent to which specific inferences are accessible to readers when reading a story silently. With this approach, researchers choose a set of theoretically important inference categories and then identify instances of those inference categories that occur in the verbal protocols. Priming targets for the priming task are derived from those instances. If priming judgments for those theoretically interesting targets are faster than those for control targets, then it is assumed that the theoretically interesting inferences are accessible to readers during online comprehension. Graesser and his colleagues have successfully used question-answering protocols to derive targets for a lexical decision task (Long & Golding, 1993; Long, Golding, & Graesser, 1993; Magliano et al., 1993; Millis & Graesser, 1994). These studies have demonstrated that the explanatory inferences exposed by why questions (e.g., causal antecedents and the goals of characters) are accessible to readers online during comprehension, whereas the associative inferences exposed by how questions (e.g., subordinate actions and instruments) and predictive inferences exposed by WHN (e.g., causal consequences and goal outcomes) are not accessible online.

Trabasso has used a recognition priming paradigm to validate the convergence between the causal network model and the inferences produced during thinking aloud (Suh & Trabasso, 1993; Trabasso & Suh, 1993). As reported earlier, Suh and Trabasso showed that readers generated the goal-based explanations during thinking aloud at the locations predicted by the causal network model. Suh and Trabasso employed a recognition priming task to demonstrate that readers generated these goal-based explanations online with a less strategic task. Suh and Trabasso had participants make recognition judgments on the first-mentioned goal statement at theoretically interesting locations in both the hierarchical and sequential versions of their stories. The causal network model predicted that the first-mentioned goal should be more highly accessible at these locations for the hierarchical versions than for the sequential version. For example, the first-mentioned goal should be highly accessible when reading the second goal because the second goal is subordinate to the first in the hierarchy, whereas the first-mentioned goal should be relatively inaccessible at the second goal statement for the sequential version because the second goal is unrelated to the first in this version (i.e., the first-mentioned goal has already been achieved). Suh and Trabasso found that the recognition judgment latencies were faster for the hierarchical versions than the sequential versions at the predicted locations. As such, Suh and Trabasso demonstrated convergence between theory, verbal protocols, and recognition priming judgments.

CONCLUSIONS

In this chapter, the three-pronged approach was discussed with an emphasis on Trabasso's influence on each prong. Trabasso's work has been particularly important in developing the theory prong and the inference elicitation prong. The causal network model provides a theoretically principled analysis for determining the causal structure inherent in a text and when causal inferences should be generated by a reader. Trabasso and Magliano (1996a) proposed think-aloud protocol analysis that allows (a) an assessment of the inferential content of the think-aloud protocols, (b) an assessment of the information source for a particular inference, and (c) an assessment of the memory operations that make that information available to a reader. The strength of the three-pronged approach is that it relies on convergence between theory and different kinds of data. The studies discussed in this chapter paint a picture of online inference processes that is consistent with the causal network model. The thoughts produced during thinking aloud, sentence reading time data, and priming data all converge in demonstrating that understanding is largely explanation based (Graesser et al., 1994; Trabasso et al., 1989). These studies also demonstrated that the causal network model can predict when different kinds of inferences will occur, such as knowledge-based explanations and associations and text-connecting inferences. Specifically, knowledge-based inferences occur primarily at sentences for which there is not a causal antecedent provided in the prior text, whereas text-connecting inferences occur when there is an antecedent in the prior text.

Tom Trabasso has recently advocated a four-pronged approach that incorporates computer simulations as a fourth prong (Langston & Trabasso, 1998; Langston et al., in press). Trabasso and his colleagues have developed a connectionist model that captures the causal relationships between text sentences, as exposed by a causal network analysis. This connectionist model produces a number of measures of online processes (i.e., number of cycles for the model to settle for each text unit after it is processed, levels of working memory activation for each text unit after it is processed, and average connection strength for each text unit after it is processed). Of the different online measures produced by the model, the average causal strength between a sentence and its predecessors accounted for a substantial amount variance in sentence reading times, sentence fit scores, the likelihood of mentioning a statement during thinking aloud, and recognition latencies for explicit text statements.

One advantage of Trabasso's four-pronged approach is that the simulation allows for an assessment of how the connectivity and accessibility of text information changes over the course of reading. Through their connectionist simulations, Langston and Trabasso (in press) demonstrated that both direct and indirect causal connections influence the accessibility and connections strengths between text ideas. As such, the entire causal history of a text representation dynamically changes over the course of reading and influences what information is accessible to the reader at any given sentence. The dynamic aspect of causal connectivity and accessibility during the course of reading would have been difficult to assess with a three-pronged approach. Specifically, a three-pronged approach incorporating the causal network model as the theory prong is particularly suited for assessing how direct causal connections between story sentences influence verbal reports and online behavioral measures. It would have be difficult to assess the impact of indirect causal connections on these two types of data without the simulation.

REFERENCES

Albrecht, J. E., & O'Brien, E. J. (1993). Updating a mental model: Maintaining both local and global coherence. *Journal of Experimental Psychology: Learning Memory, and Cognition, 19,* 1061–1070.

Bartlett, F. C. (1932). *Remembering.* Cambridge, England: Cambridge University Press.

Collins, A., Brown, J. S., & Larkin, K. M. (1980). Inference in text understanding. In R. J. Spiro, B. C. Bruce, & W. F. Brewer (Eds.), *Theoretical issues in reading comprehension,* (pp. 385–407). Hillsdale, NJ: Lawrence Erlbaum Associates.

Coté, N., & Goldman, S. R. (1998). Building representations of informational text: Evidence from children's think-aloud protocols. In H. van Oostendorp & S. R. Goldman (Eds.), *The construction of mental representations during reading.* Mahwah, NJ: Lawrence Erlbaum Associates.

Coté, N., Goldman, S. R., & Saul, E. U. (1998). Students making sense of informational text: Relations between processing and representation. *Discourse Processes, 25,* 1–53.

Ericsson, K. A., & Simon, H. A. (1994). Verbal reports as data. *Psychological Review, 87,* 215–251.

Goldman, S. R., & Varnhagen, C. K. (1986). Memory for embedded and sequential story structures. *Journal of Memory and Language, 25,* 401–418.

Graesser, A. C. (1981). *Prose comprehension beyond the word.* New York: Springer-Verlag.

Graesser, A. C., & Clark, L. F. (1985). *Structures and procedures of implicit knowledge.* Norwood, NJ: Ablex.

Graesser, A. C., Haberlandt, K., & Koizumi, D. (1987). How is reading time influenced by knowledge based inferences and world knowledge. In B. K. Britton & S. M. Glynn (Eds.), *Executive control processes in reading* (pp. 217–251). Hillsdale, NJ: Lawrence Erlbaum Associates.

Graesser, A. C., & Riha, J. R. (1984). An application of multiple regression techniques in sentence reading times. In D. Kieras & M. Just (Eds.), *New methods in comprehension research* (pp. 183–218). Hillsdale, NJ: Lawrence Erlbaum Associates.

Graesser, A. C., Singer, M., & Trabasso, T. (1994). Constructing inferences during narrative text comprehension. *Psychological Review, 101,* 371–395.

Keenan, J. M., Golding, J. M., Potts, G. R., Jennings, T. M., & Aman, C. T. (1990). Methodological issues in evaluating the occurrence of inferences. In A. C. Graesser & G. H. Bower (Eds.), *Inferences and text comprehension* (pp. 295–312). New York: Academic Press.

Keenan, J. M, Potts, G. R., Golding, J. M., & Jennings, T. M. (1990). Which elaborative inferences are drawn during reading: A question of methodologies. In D. A. Balota, G. B. Flores d'Arcais, & K. Rayner (Eds.), *Comprehension processes in reading* (pp. 377–402). Hillsdale, NJ: Lawrence Erlbaum Associates.

Langston, M. C., & Trabasso, T. (1998). Modeling causal integration and availability of information during comprehension of narrative texts. In S. R. Goldman & H. van Oostendorp (Eds.), *The construction of mental representations during reading* (pp. 29–69). Mahwah, NJ: Lawrence Erlbaum Associates.

Langston, M. C., Trabasso, T., & Magliano, J. P. (in press). Modeling online comprehension. In A. Ram & K. Moorman (Eds.), *Computational models of reading and understanding.* Cambridge, MA: MIT Press.

Long, D. L., & Bourg, T. (1996). Thinking aloud: Telling a story about a story. *Discourse Processes, 21,* 329–339.

Long, D. L., & Golding, J. M. (1993). Superordinate goal inferences: Are they automatically generated during comprehension. *Discourse Processes, 16,* 55–73.

Long, D. L., Golding, J. M., & Graesser, A. C. (1993). A test of online status of goal-related inferences. *Journal of Memory and Language, 31,* 634–647.

Lutz, M. F., & Radvansky, G. A. (1997). The fate of completed goal information in narrative comprehension. *Journal of Memory and Language, 36,* 293–310.

Mackie, J. L. (1980). *The cement of the universe.* Oxford, England: Clarendon.

Magliano, J. P., Baggett, W. B., Johnson, B. K., & Graesser, A. C. (1993). The time course in which causal antecedent and causal consequence inferences are generated. *Discourse Processes, 16,* 35–53.

Magliano, J. P., & Graesser, A. C. (1991). A three-pronged method for studying inference generation in literary text. *Poetics, 20,* 193–232.

Magliano, J. P., Trabasso, T., & LeSage, A. (1995, November). *Thinking about a movie.* Paper presented at the Psychonomics Conference, Los Angeles.

Magliano, J. P., Zwaan, R. A, & Graesser, A. C. (1998). The role of situational continuity in narrative understanding. In S. R. Goldman & H. van Oostendorp (Eds.), *The construction of mental representations during reading* (pp. 219–245). Mahwah, NJ: Lawrence Erlbaum Associates.

McKoon, G., & Ratcliff, R.(1986). Inferences about predictable events. *Journal of Experimental Psychology: Learning, Memory, and Cognition, 12,* 82–91.

McKoon, G., & Ratcliff, R. (1992). Inference during reading. *Psychological Review, 99,* 440–466.

Millis, K. K., & Graesser, A. C. (1994). The time-course of constructing knowledge-based inferences for sceintific texts. *Journal of Memory and Language, 33,* 583–599.

Millis, K. K., Morgan, D., & Graesser, A. C. (1990). The influence of knowledge-based inferences on the reading time of expository text. In A. C. Graesser & G. H. Bower (Eds.), *Inferences and text comprehension* (pp. 197–212). New York: Academic Press.

Olson, G. M., Duffy, S. A., & Mack, R. L. (1984). Thinking-out-loud as a method for studying real-time comprehension processes. In D. Kieras & M. Just (Eds.), *New methods in the study of immediate processes in comprehension* (pp. 253–286). Hillsdale, NJ: Lawrence Erlbaum Associates.

Potts, G. R., Keenan, J. M., & Golding, J. M. (1988). Assessing the occurrence of elaborative inferences: Lexical decision versus naming. *Journal of Memory and Language, 27,* 399–415.

Pressley, M., & Afflerbach, P (1995). *Verbal protocols of reading: The nature of constructively responsive reading.* Hillsdale, NJ: Lawrence Erlbaum Associates.

Stein, N. L., Folkman, S., Trabasso, T., & Richards, T. A. (1997). Appraisal and goal processes as predictors of psychological well-being in bereaved caregivers. *Journal of Personality and Social Psychology, 72,* 872–884.

Stein, N. L., & Glenn, C. G. (1979). An analysis of story comprehension in elementary school children. In R. O. Freedle (Ed.), *New directions in discourse processing* (Vol. 2, pp. 53–120). Hilldale, NJ: Lawrence Erlbaum Associates.

Stein, N. L., & Trabasso, T. (1985). The search after meaning: Comprehension and comprehension monitoring. In F. J. Morrison, C. Lord, & D. Keating (Eds.), *Applied developmental psychology* (Vol. 2, pp. 33–58). San Diego: Academic Press.

Suh, S., & Trabasso, T. (1993). Inferences during reading: Converging evidence from discourse analysis, talk-aloud protocols, and recognition priming. *Journal of Memory and Language, 32,* 279–301.

Trabasso, T., & Magliano, J. P. (1996a). Conscious understanding during text comprehension. *Discourse Processes, 21,* 255–288.

Trabasso, T., & Magliano, J. P. (1996b). How do children understand what they read and what can we do to help them? In M Graves, P. van den Broek, & B. Taylor (Eds.), *The first R: a right of all children* (pp. 160–188). New York: Columbia University Press.

Trabasso, T., & Sperry, L. (1985). Causal relatedness and importance of story events. *Journal of Memory and Language, 24,* 595–611.

Trabasso, T., & Suh, S. (1993). Understanding text: Achieving explanatory coherence through online inferences and mental operations in working memory. *Discourse Processes, 16,* 3–34.

Trabasso, T., van den Broek, P., & Suh, S. (1989). Logical necessity and transitivity of causal relations in the representation of stories. *Discourse Processes, 12,* 1–25.

Trabasso, T., & van den Broek, P. (1985). Causal thinking and the representation of narrative events. *Journal of Memory and Language, 24,* 612–630.

van den Broek, P. (1989). The effects of causal relations and hierarchical position on the importance of story statements. *Journal of Memory and Language, 27,* 1–22.

van den Broek, P. W. (1990). The causal inference maker: Towards a process model of inference generation in text comprehension. In D. A. Balota, G. B. Flores d'Arcais, & K. Rayner (Eds.), *Comprehension processes in reading* (pp. 423–446). Hillsdale, NJ: Lawrence Erlbaum Associates.

van den Broek, P. (1995). Comprehension and memory of narrative text: Inferences and coherence. In M. A. Gernsbacher (Ed.), *Handbook of psycholinguistics* (pp. 539–588). New York: Academic Press.

van den Broek, P., & Lorch, R. F. (1993). Network representations of causal relations in memory for narrative texts: Evidence from primed recognition. *Discourse Processes, 16,* 75–98.

van den Broek, P., Rohleder, L., & Narváez, D. (1996). Causal inferences and the comprehension of literary text. In R. J. Kreuz & M. S. MacNealy (Eds.), *Empirical approaches to literature and aesthetics* (pp 201– 220). Norwood, NJ: Ablex.

Whitney, P., & Budd, D. (1996). Thinking-aloud protocols and the study of comprehension. *Discourse Processes, 21,* 341–351.

Whitney, P., & Williams-Whitney, D. (1990). Toward a contextualist view of elaborative inferences. In A. C. Graesser & G. H. Bower (Eds.), *Inferences and text comprehension* (pp. 279–293). New York: Academic Press.

Zwaan, R. A., & Brown, C. M. (1996). The influence of language proficiency and comprehension skill on situation-model construction. *Discourse Processes, 21,* 289–328.

6

Situation Models and Concepts in Story Comprehension

Arthur C. Graesser and Katja Wiemer-Hastings
The University of Memphis

When a reader comprehends a story, the reader constructs a mental microworld. The microworld includes the core plot that sustains the interest of the reader: the characters who perform actions in pursuit of goals, events that present obstacles to goals, conflicts between characters, clever methods of resolving conflicts, and emotional reactions to events and conflicts. The mental microworld sometimes includes elaborations that flesh out the plot and add color: the spatial setting, the style and procedure of actions, props, objects, properties of objects, and traits of agents. The microworld might also have content that refers to the mental states of the characters (i.e., what they believe, know, see, want, and feel) and the pragmatic agents who tell the story (i.e., the imaginary narrator and the camera operator). Discourse psychologists have used the term *situation model* when they refer to this mental microworld that is constructed by the comprehender of a story (Graesser, Millis, & Zwaan, 1997; Kintsch, 1998).

Discourse psychologists are in harmony about some general beliefs about the construction of situation models for stories. They uniformly agree that the explicit text does not sufficiently constrain and determine what information goes into the situation model. Instead, much of the information is actively constructed by tapping world knowledge and by inferentially filling in gaps in the explicit text (Bower, Black, & Turner, 1979; Graesser, Millis, et al., 1997; Kintsch, 1998; Trabasso & Magliano, 1996; van den Broek, 1994). This claim is perfectly obvious and uncontroversial. But the claim has nontrivial implications because it forces us to discard any theory, field, or methodology that attempts to derive mental representations solely from surface code, language, discourse markers, and rhetorical patterns. Discourse psychologists agree that the situation model is easy to construct when it has *verisimilitude*; that is, it is believable from the standpoint of everyday experiences and knowledge about the world. But they also believe that an interesting story has some deviation from normalcy (Graesser, Kassler, Kreuz, & McLain-Allen, 1998; Miall & Kuiken, 1994; Schank & Abelson, 1996). A story would fail to hold the attention of an adult if it perfectly meshed with world knowledge and mundane experiences.

As any scientist would expect, however, discourse psychologists disagree about the exact processes and products of situation model construction. What processing mechanisms systematically generate the microworlds? What content ends up being encoded? What inferences emerge? These challenging questions have attracted the attention of Tom Trabasso and his collaborators in discourse psychology for many years (Graesser, Singer, & Trabasso, 1994; Singer, Graesser, & Trabasso, 1994; Stein, Trabasso, & Liwag, 1993; Suh & Trabasso, 1993; Trabasso & Magliano, 1996; Trabasso & van den Broek, 1985). At this point in the science, after approximately 20 years of research, there are several theories and models of situation model construction that make different claims about processes and representations. These include the structure-building framework (Gernsbacher, 1997), the construction-integration model (Kintsch, 1998), the constructionist theory (Graesser et al., 1994), the event-indexing model (Zwaan, Langston, & Graesser, 1995; Zwaan, Magliano, & Graesser, 1995; Zwaan & Radvansky, 1998), the landscape model (van den Broek, Young, Tzeng, & Linderholm, 1999), the resonance model (Myers, O'Brien, Albrecht, & Mason, 1994; O'Brien, Raney, Albrecht, & Rayner, 1997), and hybrid models (Goldman, Varma, & Coté, 1996). The available data and theories constrain our beliefs about plausible psychological mechanisms, but there is no mature theory with widespread acceptance.

In this chapter, we hope to accomplish two goals. Our first goal is to specify some predictions of a constructionist theory that we have pursued in recent years (Graesser et al., 1994). This constructionist theory makes some discriminating commitments about the classes of inferences that are versus are not generated during story comprehension. The predictions of the constructionist theory directly apply to texts in the narrative genre, but we are convinced that most of the predictions would also hold up in investigations of expository texts on history (Perfetti, Britt, & Georgi, 1995; Voss & Silfies, 1996) and even scientific mechanisms (Coté, Goldman, & Saul, 1998; Goldman et al., 1996; Graesser & Bertus, 1998; Millis &

Graesser, 1994; Singer, Harkness, & Stewart, 1997). Our second goal is to clarify the role of generic concepts in situation model construction. Generic concepts contain most of the world knowledge that furnishes the inferences, as well as the interpretations of explicit text (Graesser & Clark, 1985). Therefore, it is important to explain how the knowledge in generic packages of information gets passed to the specific situation model that a reader creates for a particular story.

BUILDING SITUATION MODELS FOR STORIES: A CONSTRUCTIONIST THEORY

The constructionist theory proposed by Graesser et al. (1994) accounts for many empirical facts about the construction of situation models during story comprehension. Some of the assumptions of this theory are not particularly controversial because they are compatible with alternative models. For example, most models in discourse psychology assume that there are multiple levels of representation (lexical, syntactic, textbase, situation model), multiple information sources (text, generic knowledge, episodic experiences), multiple memory stores (short-term memory, working memory, long-term memory), constraint satisfaction, and a discourse focus. As with most models in discourse psychology, the constructionist theory assumes that the inferences and content of the situation model are systematically influenced by variations in reader goals. Articulating such assumptions is pretty much business as usual in the discourse-processing enterprise. There are two distinctive assumptions of the constructionist theory that systematically restrict the content that ends up being constructed in the situation model: the *coherence* assumption and the *explanation* assumption. These assumptions of our constructionist theory substantially narrow down the set of inferences that are routinely generated online during story comprehension.

According to the coherence assumption, the comprehender attempts to build a situation model that establishes both local and global coherence among the actions, events, and states that are explicitly mentioned in the text. Local coherence is established if an incoming explicit statement (S) can be linked conceptually to a recent proposition (P) that resides in working memory (WM). It appears that comprehenders monitor several dimensions of coherence simultaneously during these attempts at achieving local coherence (i.e., linking S to a P in working memory). Research by Zwaan

(Zwaan, Magliano, & Graesser, 1995; Zwaan & Radvansky, 1998; see also Zwaan, chap. 7, this volume) has reported that reading time increases for a sentence in a story if it involves a break in coherence (or continuity) on any one of the following five dimensions:

1. *Protagonist.* The protagonists in explicit statement S are not among the protagonists in WM.
2. *Temporality.* The action or event expressed in S involves a gap or shift in the chronological timeline.
3. *Causality.* The action or event expressed in S does not causally flow from the content in WM.
4. *Motivation.* The action expressed in S is not part of an agent's plan in WM.
5. *Spatiality.* The action or event expressed in S is in a different spatial region from the content in WM.

It should be noted that an increase in reading time would still occur when there is continuity on four out of the five dimensions. That is, a break on any one dimension is sufficient to increase reading time. Moreover, reading time is progressively longer to the extent that there are continuity breaks on more dimensions.

Global coherence is established when local chunks of information can be organized into higher order chunks. For example, several sentences in the text might funnel into a chunk that signifies an argument between a husband and wife over money, whereas the next stretch of text signifies a physical altercation and the final stretch of text signifies a divorce. These three chunks might be coherently organized by a moral, main point, or theme such as "money is the root of all evil." A good theme has tentacles to many chunks, which in turn have tentacles to many episodes in the text (Lehnert, 1981). In order to establish global coherence, the incoming explicit statement S sometimes needs to be linked to content much earlier in the text, content that is no longer in WM. So an interesting research question arises as to how well readers can achieve the global coherence. If a reader is comprehending statement S on page 143 in a novel, can the reader reinstate relevant content from page 34? If the story is well written and considerate, then global coherence presumably can be successfully achieved. However, all bets are off when the story is pointless, choppy, and difficult to relate to general world knowledge. The reader settles for local coherence under these adverse conditions. Unfortunately, the stories in many psychology experiments have leaned toward the inconsiderate end of the continuum, so it has been difficult to assess how well readers can achieve global coherence.

According to the explanation assumption, comprehenders attempt to explain *why* the explicit actions, events, and states occur and *why* the author bothers to mention information in the text. Comprehension is driven by why questions to a much greater extent than other types of questions (when, where, how, what-happens-next). The causal explanation of an action A includes the motives (superordinate goals) for performing A and the events/states that initiate these motives. The causal explanation of an event E includes the causal antecedents that lead up to E. A causal explanation of an unusual, nonstandard, or unexpected constituent in a text (e.g., word, phrases, clause, sentence) would attempt to explain why an author bothered to mention something or why it was expressed in an atypical wording or style. In order to illustrate these types of explanations, consider the following scenario that might occur in a garden-variety romantic story:

> Two brothers were always in competition over money and success. George unfortunately went bankrupt the same year that Bill made his first million. Bill threw a New Year's Eve party that ended up being quite a festive occasion. But everyone sobered up very quickly when George was caught kissing Bill's wife.

Listed next are a set of potential inferences at the point in the story when George is caught kissing Bill's wife. Each inference is evaluated from the standpoint of being constructed in the situation model:

1. *George wanted to hurt Bill.* This is a superordinate goal of the act of George kissing Bill's wife. Superordinate goals explain *why* acts occur, so this inference would have a high likelihood of being constructed in the situation model.
2. *George touched the cheek of Bill's wife.* This elaborates the subplan (i.e., subgoal, procedure, method, style) of the act of George kissing Bill's wife. This information specifies *how* the kissing occurred, but does not explain *why* it occurred, so it would have a low likelihood of being constructed in the situation model. It is mere ornamentation, not part of the explanatory guts of the plot.
3. *George was jealous of Bill.* This is an important causal antecedent that leads to the kiss. George's bankruptcy and Bill's financial success caused George to be jealous of Bill, which in turn triggered the goal of George wanting to hurt Bill. This chain of causal antecedents to the kiss explains *why* the kiss occurred, so it would be constructed in the situation model.
4. *Bill slugged George.* This is a plausible causal consequence (expectation) of George's kiss. It answers a what-happens-next question, but not a why question, so it is probably not generated online in the situation model.

5. *George was in the bedroom.* This is a plausible spatial setting for a kiss. But it fails to explain the kiss, so it is not likely to be constructed in the situation model.

6. *George's kiss was a romantic kiss.* The kiss was hardly a gentle peck of friendship in the spirit of the New Year. A gentle peck would not explain *why* the author used the words *But* and *was caught,* and *why* the author bothered mentioning that the guests quickly sobered up. The romantic kiss inference is constructed online because it explains why the author bothers mentioning particular words, expressions, and statements.

These examples illustrate that the constructionist theory provides a principled and decisive foundation for predicting what inferences are built in the situation model during comprehension. The comprehender generates inferences selectively and strategically. The explanation-based inferences and coherence-based inferences are products of an intentional comprehension strategy (either conscious or unconscious) that attempts to explain the text and to integrate the content of the text coherently. Of course, there may be other inferences that are products of passive activations of knowledge, such as an inference that is activated by multiple information sources (Kintsch, 1998) or an inference that is very typical of a package of relevant world knowledge (Bower et al., 1979; Schank & Abelson, 1996). The constructionist theory has a separate set of assumptions to handle inferences that are products of these passive activations (which are discussed a bit more in the last section of this chapter). These passive-activation inferences are generated by most theoretical models, so they are not unique to the constructionist theory. The distinctive claim of the constructionist theory is that readers intentionally attempt to achieve explanation and coherence, and that particular classes of inference are products of these intentions. Moreover, these intentions are well weathered and automatized in the comprehension mechanisms. These intentions are nearly invariant among readers, text genres, and reading situations. When we observe the world for 16 hours per day in everyday life, we also intentionally construct coherent explanations of what we see and what we do.

As with any theory, it would be appropriate to precisely specify the limits of the theory and offer a list of hedges. For example, the explanation-based and coherence-based inferences might not be as consistently generated when the reader is reading very quickly, when working-memory capacity is exceeded, when the reader is unconvinced the author is coherent, or when the reader has an unusual reading goal (e.g., monitoring the spatial location of an object). According to the theory's *reader goal assumption,* readers attempt to generate other classes of inferences when they have particular goals that monitor such inferences. Graesser et al. (1994) enumerated the various conditions, limitations, and augmentations of the theory, but for the present purposes we sketch the simple picture.

There is substantial evidence for the distinctive predictions of the constructionist theory. It is beyond the scope of this chapter to review all of this evidence, but we point out two forms of evidence. One type of empirical evidence comes from studies that collect *think-aloud* protocols from readers as they comprehend stories, sentence by sentence. The think-aloud protocols provide a snapshot of the approximate conscious content of the reader at different points in the story. Most of the content in these think-aloud protocols for stories consists of knowledge-based inferences, as opposed to the explicit content (Trabasso & Magliano, 1996; Zwaan & Brown, 1996). The studies by Trabasso and Zwaan have also revealed that *explanation*-based inferences account for most of the content and are far more plentiful than *expectations* about future plot and nonexplanatory *associations.*

A second type of empirical evidence comes from studies that collect reaction time data (such as word naming and lexical decision latencies) for test strings that appear immediately after sentences in the text. The test words match inferences in different theoretical categories, such as superordinate goals, subordinate goals, causal antecedents, causal consequences, spatial setting, and so on. Researchers can impose careful control over study time by using a rapid serial visual presentation task, with each word being presented for a short duration (250–500 milliseconds per word). Researchers can trace the time course of inference activation and encoding by presenting test words at specific durations after the end of the sentence (e.g., 150, 500, 1,000 vs. 2,000 milliseconds). The extent to which an inference is encoded can be measured by subtracting the latency for a word in an inference condition from the latency for the same word in an unrelated text condition. These well-controlled experiments have shown patterns that fit the predictions of the constructionist theory. Inference activation scores are higher for the superordinate goals of an action sentence than for subordinate goals (Long, Golding, & Graesser, 1992), and higher for the causal antecedents of an event sentence than for causal consequences (Magliano, Baggett, Johnson, & Graesser, 1993).

Spatial inferences have traditionally received a large amount of attention in the literature on situation models (Glenberg, Meyer, & Lindem, 1987; Morrow, Greenspan, & Bower, 1987), so the question arises what the constructionist theory claims about this class of inference. According to the constructionist theory, the comprehender does construct a *spatial setting* for the story episodes, but the representation is stark and fuzzy unless there is substantial support from explicit information and auxiliary information sources. The mental representation of the spatial setting can be quite elaborate if (a) it is described explicitly in the text, (b) the reader has prior experience with the setting (e.g., a story about Memphis for us coauthors), (c) the reader has memorized a map of the setting, or (d) the reader has a special goal of tracking spatiality at a fine-grained level. Otherwise, the setting is a bare-bones, stark, and fuzzy frame rather than a rich sensuous image (which would be easy to capture on film). The mental director constructs an abstract fuzzy stage for each spatial region or setting mentioned in the story. The spatial setting for the example text would be an abstract house, with no details about the floorplan, location of characters, location of objects, and visual properties of entities. George's kiss would occur somewhere in the house, but there would be no details about which room, the relative location of Bill, George, and Bill's wife, or the relative height of Bill and George. The constructionist theory assumes there is a rudimentary topological organization of global spatial regions, but no details about layout of entities and visual features unless they are explicitly mentioned in the text, are strongly activated by passive constraint satisfaction mechanisms, or are intentionally tracked by the reader.

Surface features of the text sometimes provide "processing instructions" (Givon, 1993) for constructing content that otherwise would not be built according to the constructionist theory. One example of this consists of verb tense and aspect. The normal tense and aspect of the verb is *past* tense and *completed* aspect when a sequence of episodes in the plot is described in a story (Hopper, 1979). The narrator is speaking to the narratee "now," but referring to a chronological sequence of episodes in the past. The aspect of actions and events is completed (e.g., achieved, finished) rather than in progress (e.g., beginning, in the midst of the process). The following sequence of episodes would fit the normal way of articulating episodes in the story: "George kissed Bill's wife. Bill threw George out of his house." In this canonical articulation, the kissing is completed, the throwing out is completed, and the kissing precedes the throwing out on the chronological timeline. However, speakers sometimes deviate from this canonical use of time and aspect by using a past progressive form (George was kissing Bill's wife, George started kissing Bill's wife) or a present progressive form (George is kissing Bill's wife). This deviation from the canonical form draws the readers attention to the process of the act rather than the achievement of the act, so more of the subplans, subgoals, processes, procedures, and style specifications should be constructed in the situation model. As discussed earlier, these frivolous elaborations should not normally be included in the situation model according to the constructionist theory.

However, the construction of such inferences is in fact compatible with the constructionist theory by virtue of the explanation assumption: The reader attempts to explain *why* the author used the unusual tense and aspect that deviates from the normal canonical form. The obvious explanation is that the author wants the reader to construct content that refers to the process or lack of completion or the action or event.

The constructionist theory therefore does embrace the view that many surface forms have a discourse function that directs the construction of the situation model. Our theory also provides the more distinctive claim that many of these functions are linked to explanations: The comprehender attempts to explain why the author uses particular words, syntax, and semantic expressions (presumably unconsciously). Of course, some readers have a fine palate and some a blunt palate when they build such explanations that address pragmatics and language. These inferences are perhaps routinely made only by readers who have sufficient expertise in language and literature (Graesser et al., 1998).

The constructionist theory that was originally proposed by Graesser et al. (1994) acknowledged the existence of *pragmatic* agents, but there were no specific commitments about what characteristics of these agents are constructed in the situation model. Pragmatic agents refer to the parties of the communication when stories are told. The pragmatic agents include the speaker, addressee, and side participant in oral conversation (Clark, 1996; Keysar, 1994), and the narrator, narratee, author, and reader when stories are read (Graesser, Swamer, & Hu, 1997; Keysar, 1994). The critical question, from the standpoint of this chapter, is whether the reader constructs these pragmatic agents in the situation model during the course of reading stories. For example, does the reader construct an imaginary narrator who tells the story and who manipulates the characters, objects, and props on the stage of the mental microworld? According to the constructionist theory, the reader does not build these agents unless there are noncanonical features in the text that signal such agents. A contrast between third-person narration and first-person narration illustrates what is being proposed. A third-person narrator is the canonical perspective when stories are written, as in the example story involving George, Bill, and the wife (e.g., "George was caught kissing Bill's wife"). The narrator is invisible in the sense that there are no features in the text that explicitly refer to the narrator. The narrator is omniscient when the narrator is all-knowing, can observe all of the episodes that occur, and has privileged access to the minds of all characters. The third-person narrator is *not* theoretically constructed because there are no unusual surface characteristics to explain. In contrast, the narrator is constructed when there are deviations from the canonical form, as in the case of first-person narration (e.g., "I was caught kissing Bill's wife," where the narrator is amalgamated with the character George) and second-person narration (e.g., "You are caught kissing Bill's wife," an amalgamation of the character George, the narrator, the narratee, and the reader). The reader attempts to explain these deviations from canonical form by constructing a narrator that takes the point of view of particular characters. Available research is compatible with the claim that first-person narrators are salient to the comprehender, whereas third-person narrators lean toward the invisible end of the continuum (Duchan, Bruder, & Hewitt, 1995; Graesser, Bowers, Olde, & Pomeroy, 1997).

In summary, the constructionist theory does make discriminating claims about what content gets constructed in the situation model during story comprehension. There is a pervasive comprehension strategy that constructs those inferences that provide explanations and coherence. These are the signature inferences of the constructionist theory. Additional inferences are constructed by virtue of particular reader goals and passive activations of knowledge from multiple information sources; such inferences would be predicted by most models in discourse psychology, so they are not unique to the constructionist theory.

THE ROLE OF GENERIC CONCEPTS IN THE CONSTRUCTION OF INFERENCES IN SITUATION MODELS

Knowledge-based inferences come from somewhere of course. A mature theory of inference generation needs to specify the information sources that supply the inferences in the situation model. A mature theory needs to specify the processes that determine what content in an information source does versus does not end up being an inference. This section explores the information sources that furnish the inferences in the situation model.

One possible mechanism is that inferences are passed down from the generic concepts that get activated during story comprehension. For example, the sentence "George kissed Bill's wife" would activate the generic concepts that are affiliated with the explicit content words: MAN, KISSING, WIFE. In addition to these *word-activated* generic concepts, *pattern-activated* generic concepts are activated by sets of words, contextual features, and compound cues (e.g., BETRAYAL, INFIDELITY). Each generic concept is a rich package of structured information that contains hundreds, if not thousands, of conceptual elements and relations. Much of this rich content is automatically activated when the generic content is activated. The content in these generic concepts presumably supplies many of the inferences that are constructed when a particular sentence is comprehended in a particular passage.

Some of the knowledge-based inferences match information that is stored in the generic concepts. These *generic-concept* inferences come in two varieties: *direct matches* and *argument substitution* matches (Graesser & Clark, 1985). In the case of direct matches, the content of the inference about the text is a direct match with the content in the generic concept. For example, the instrumental inference that "the kissing was with the lips" would perfectly match a proposition node in the generic concept of KISSING. In the case of argument substitution matches, a specific concrete argument in the inference would be substituted with a semantically depleted argument of a proposition in the generic concept. For example, the emotional state inference "George liked the wife" would match the proposition "X likes Y" in the generic concept of KISSING. George would correspond to X, and wife would correspond to Y.

It is conceivable that the generic concept inferences do not go the entire distance in accounting for the inferences associated with a particular sentence. There might be unique characteristics of the situation model that do not match any of the propositions in the generic concepts. We refer to these inferences as novel-situational inferences. For example, when the reader comprehends the sentence "But everyone sobered up very quickly when George was caught kissing Bill's wife," the following spatial inferences are potential *novel-situational* inferences because they would not be inherited by any of the generic concepts: "The guests (everyone) were in the same room as George," "Bill was in the same room as George," or even "George was in the same room as Bill's wife." These unique spatial constructions are determined by the idiosyncratic properties of the mental stage of the situation model, not by the content inherited from generic concepts. Novel-situational inferences presumably demand many more processing resources to construct and are therefore at the mercy of the reader's comprehension strategies. A simple, but plausible, generalization is that generic-concept inferences are automatic whereas novel-situational inferences are strategic and time consuming.

Graesser and Clark (1985) reported some analyses that assessed the extent to which the knowledge-based inferences in stories are novel-situational versus generic-concept inferences. Word-activated and pattern-activated concepts were identified in four short stories of approximately 150 words (or 20 propositions). A pool of inferences was collected from verbal protocols that questioned readers about each explicit action, event, and state in the text. That is, after a reader comprehended statement S, the reader answered a why, how, and what-happened-next (WHN) question. Approximately 250 individual inferences were extracted from a passage when we used this question-answering methodology to expose inferences. The issue under consideration was how many of these inferences were novel-situational versus generic-concept inferences.

Graesser and Clark (1985) used knowledge-engineering methods to extract content from each of the generic concepts associated with a story. The knowledge extraction method had two steps. College students first listed properties, traits, behaviors, and events for each generic concept. This "free-generation" method has been frequently used to extract content about concepts (see Barsalou, 1993; Tversky & Hemenway, 1984). All statements listed by two or more participants were submitted to a second step in which question-answering protocols were collected. That is, a why, how, and WHN question was asked about each statement that had been generated by two or more participants in the free-generation task. This question-answering step produced many additional statements. All statements generated by two or more participants (in either the free-generation task or the question-answering task) were included in the knowledge package associated with a generic concept. An average generic concept had 166 individual statements that were extracted from this "free-generation plus question-answering" method of knowledge extraction.

Graesser and Clark (1985) discovered that the vast majority of inferences in the stories were generic-concept inferences. They found that 72% of the story inferences matched one or more generic concepts in working memory; 63% of the inferences came from a word-activated concept, whereas an additional 9% came from exclusively pattern-activated concepts. Therefore, the generic concepts associated with explicit words in the text supplied the majority of the knowledge-based inferences. The remaining 28% of the story inferences were novel-situational inferences.

There was one additional finding reported in the Graesser and Clark (1985) study that is particularly noteworthy from the standpoint of the constructionist theory. Most of the novel-situational inferences were the frivolous elaborative inferences that the constructionist theory predicts are *not* generated online during comprehension. The majority of the novel inferences were either subgoals, subprocedures, style specifications, spatial inferences, and other descriptions that require deictic grounding in a specific situation. If it truly is the case that these elaborative and deictic inferences are not generated online during comprehension, then a very strong conclusion can be mined from the data. The strong conclusion is that virtually all of the inferences that are constructed online during story comprehension are inherited from the generic concepts associated with a story (particularly those captured by explicit words). If this conclusion is correct, then the central challenge for researchers is to identify what propositions in the generic concepts are versus are not passed down to the situation model, and also what mechanisms explain this convergence. Graesser and Clark identified some of these convergence mechanisms, but it is beyond the scope of this chapter to address these.

The preceding results underscore the importance of generic concepts in studies of discourse comprehension. For some unfortunate reasons, theories of concept representation (Barsalou, 1993; Smith & Medin, 1981) have traditionally not been closely integrated with theories in discourse psychology. The dissociation between these two fertile research areas can perhaps be attributed to a historical accident rather than a conscious effort to preserve their insulation. The most recent models in discourse psychology have been more prone to embrace theories of concepts (Goldman et al., 1996; Kintsch, 1998), so we can perhaps expect more cross-talk in the future. It is quite apparent that generic concepts need to be tapped when the comprehender interprets explicit text, formulates relations between text constituents, and generates knowledge-based inferences. A model without generic concepts in discourse psychology is entirely bankrupt.

Theories of concept representation may also benefit from the models in discourse psychology. We know that concepts evolve from episodic memories (Nelson, 1986; Smith & Medin, 1981), and we know that much of episodic memory is embodied in narrative and discourse (Kintsch, 1998; Schank & Abelson, 1996; Stein et al., 1993; Trabasso & van den Broek, 1985). It should therefore be perfectly obvious that narrative and discourse critically shape our representation of concepts. Much of the narrative and discourse no doubt recedes as the episodic memories linked to a concept fade. But some of these narratives seem to be permanently etched on concept representations. What would the U.S. flag be without the Revolutionary War? What would the balalaika be without Dr. Zhivago?

In closing, the process of constructing the microworlds of stories is not as mysterious as it was 20 years ago. The representations and knowledge-based inferences are systematically constructed according to plausible cognitive mechanisms. Thus, when it comes to delineating the inferences in situation models for stories, it is inappropriate to conclude that "anything goes" or that "nothing goes." The constructionist theory of Graesser et al.

(1994) provides one account of the systematicity. Readers reliably generate inferences that explain the text and that provide local and global coherence. Readers also construct inferences that are relevant to their comprehension goals or are products of passive activations of knowledge. The vast majority of knowledge-based inferences are passed down from generic concepts that get activated by content words and patterns of information. In contrast, novel-situational inferences are considerably less frequent and require more cognitive resources to construct. It remains to be seen how well this constructionist theory survives the tests of time, empirical data, and alternative theories.

ACKNOWLEDGMENTS

Correspondence concerning this article should be addressed to Arthur C. Graesser, Department of Psychology, Campus Box 526400, The University of Memphis, Memphis, TN 38152-6400, a-graesser@memphis.edu. This research was partially funded by grants to the first author by the Office of Naval Research (N00014-95-11113 and N00014-98-1-0331) and the National Science Foundation (SBR 9720314).

REFERENCES

Barsalou, L. W. (1993). Flexibility, structure, and linguistic vagary in concepts: Manifestations of a compositional system of perceptual symbols. In A. C. Collins, S. E. Gathercole, & M. A. Conway (Eds.), *Theories of memories* (pp. 29–101). Hovey, England: Lawrence Erlbaum Associates.

Bower, G. H., Black, J. B., & Turner, T. J. (1979). Scripts in memory for text. *Cognitive Psychology, 11,* 177–220.

Clark, H. H. (1996). *Using language.* Cambridge, England: Cambridge University Press.

Coté, N., Goldman, S. R., & Saul, E. U. (1998). Students making sense of informational text: Relations between processing and representation. *Discourse Processes, 25,* 1–53.

Duchan, J. F., Bruder, G. A., & Hewitt, L. E. (Eds.). (1995). *Deixis in narrative: A cognitive science perspective.* Hillsdale, NJ: Lawrence Erlbaum Associates.

Gernsbacher, M. A. (1997). Two decades of structure building. *Discourse Processes, 23,* 265–304.

Givon, T. (1993). Coherence in text, coherence in mind. *Pragmatics & Cognition, 1,* 171–227.

Glenberg, A. M., Meyer, M., & Lindem, K. (1987). Mental models contribute to foregrounding during text comprehension. *Journal of Memory and Language, 26,* 69–83.

Goldman, S. R., Varma, S., & Coté, N. (1996). Extending capacity-constrained construction-integration: Toward "smarter" and flexible models of text comprehension. In B. K. Britton & A. C. Graesser (Eds.), *Models of understanding text* (pp. 73–114). Mahwah, NJ: Lawrence Erlbaum Associates.

Graesser, A. C., & Bertus, E. (1998). The construction of causal inferences while reading expository texts on science and technology. *Journal of the Scientific Studies of Reading, 2,* 247–269.

Graesser, A. C., Bowers, C. A., Olde, B., & Pomeroy, V. (1997). *Who said what? Source memory for narrator and character agents in literary stories.* Unpublished manuscript, The University of Memphis, Memphis.

Graesser, A. C., & Clark, L. C. (1985). *Structures and procedures of implicit knowledge.* Norwood, NJ: Ablex.

Graesser, A. C., Kassler, M. A., Kreuz, R. J., & McLain-Allen, B. (1998). Verification of statements about story worlds that deviate from normal conceptions of time: What is true about Einstein's Dreams? *Cognitive Psychology, 35,* 246–301.

Graesser, A. C., Millis, K. K., & Zwaan. R. A. (1997). Discourse comprehension. *Annual Review of Psychology, 48,* 163–189.

Graesser, A. C., Singer, M., & Trabasso, T. (1994). Constructing inferences during narrative text comprehension. *Psychological Review, 101,* 371–395.

Graesser, A. C., Swamer, S., & Hu, X. (1997). Quantitative discourse psychology. *Discourse Processes, 23,* 229–264.

Hopper, P. J. (1979). Aspect and foregrounding in discourse. In T. Givon (Ed.), *Syntax and Semantics* (Vol. 12, pp. 213–241). New York: Academic Press.

Keysar, B. (1994). The illusory transparency of intention: Linguistic perspective taking in text. *Cognitive Psychology, 26,* 165–208.

Kintsch, W. (1998). *Comprehension: A paradigm for cognition.* Cambridge, MA: Cambridge University Press.

Lehnert, W. G. (1981). Plot units and narrative summarization. *Cognitive Science, 5,* 293–331.

Long, D. L., Golding, J. M. & Graesser, A. C. (1992). Test of the on-line status of goal-related inferences. *Journal of Memory and Language, 31,* 634–647.

Magliano, J. P., Baggett, W. B., Johnson, B. K., & Graesser, A. C. (1993). The time course of generating causal antecedent and causal consequence inferences. *Discourse Processes, 16,* 35–53.

Miall, D. S., & Kuiken, D. (1994). Beyond text theory: Understanding literary response. *Discourse Processes, 17,* 337–352.

Millis, K., & Graesser, A. C. (1994). The time-course of constructing knowledge-based inferences for scientific texts. *Journal of Memory and Language, 33,* 583–599.

Morrow, D. G., Greenspan, S. L., & Bower, G. H. (1987). Accessibility and situation models in narrative comprehension. *Journal of Memory and Language, 26,* 165–187.

Myers, J. L., O'Brien, E. J., Albrecht, J. E., & Mason, R. A. (1994). Maintaining global coherence during reading. *Journal of Experimental Psychology: Human Learning, Memory, and Cognition, 20,* 876–886.

Nelson, K. (1986). *Event knowledge: Structure and function in development.* Hillsdale, NJ: Lawrence Erlbaum Associates.

O'Brien, E. J., Raney, G. E., Albrecht, J. E., & Rayner, K. (1997). Processes involved in the resolution of explicit anaphors. *Discourse Processes, 23,* 1–24.

Perfetti, C. A., Britt, M. A., & Georgi, M. C. (1995). *Text-based learning and reasoning: Studies in history.* Hillsdale, NJ: Lawrence Erlbaum Associates.

Schank, R. C., & Abelson, R. P. (1995). Knowledge and memory: The real story. In R. S. Wyer (Ed.), Knowledge and memory: The real story (pp. 1–86). *Series: Advances in social cognition.* Mahwah, NJ: Lawrence Erlbaum Associates.

Singer, M., Graesser, A. C., & Trabasso, T. (1994). Minimal or global inference during reading. *Journal of Memory and Language, 33,* 421–441.

Singer, M., Harkness, D., & Stewart, S. T. (1997). Constructing inferences in expository text comprehension. *Discourse Processes, 24,* 199–228.

Smith, E. E., & Medin, D. L. (1981). *Categories and concepts.* Cambridge, MA: Harvard University Press.

Stein, N. L., Trabasso, T., & Liwag, M. D . (1993). The representation and organization of emotional experience: Unfolding the emotional episode. In M. L. Lewis & J. Haviland (Eds.), *Handbook of emotions* (pp. 279–300). New York: Guilford.

Suh, S. Y., & Trabasso, T. (1993). Inferences during reading: Converging evidence from discourse analysis, talk-aloud protocols, and recognition priming. *Journal of Memory and Language, 32,* 279–300.

Trabasso, T., & Magliano, J. P. (1996). Conscious understanding during comprehension. *Discourse Processes, 21,* 255–287.

Trabasso, T., & van den Broek, P. (1985). Causal thinking and the representation of narrative events. *Journal of Memory and Language, 24,* 612–630.

Tversky, B., & Hemenway, K. (1984). Objects, parts, and categories. *Journal of Experimental Psychology: General, 113,* 169–193.

van den Broek, P. (1994). Comprehension and memory of narrative texts: Inferences and coherence. In M. A. Gernsbacher (Ed.), *Handbook of psycholinguistics* (pp. 539–588). New York: Academic Press.

van den Broek, P., Young, M., Tzeng, Y., & Linderholm, T. (1999). The landscape model of reading inferences and the on-line construction of memory representations. In H. van Oostendorp & S. R. Goldman (Eds.), *The construction of mental representations during reading* (pp. 71–98). Mahwah, NJ: Lawrence Erlbaum Associates.

Voss, J. F., & Silfies, L. N. (1996). Learning from history text: The interaction of knowledge and comprehension skill with text structure. *Cognition & Instruction, 14,* 45–68.

Zwaan, R. A., & Brown, C. M. (1996). The influence of language proficiency and comprehension skill on situation model construction. *Discourse Processes, 21,* 289–327.

Zwaan, R. A., Langston, M. C., & Graesser, A. C. (1995). The construction of situation models in narrative comprehension: An event-indexing model. *Psychological Science, 6,* 292–297.

Zwaan, R. A., Magliano, J. P., & Graesser, A. C. (1995). Dimensions of situation model construction in narrative comprehension. *Journal of Experimental Psychology: Learning, Memory, and Cognition, 21,* 386–397.

Zwaan, R. A., & Radvansky, G. A. (1998). Situation models in language comprehension and memory. *Psychological Bulletin, 123,* 162–185.

7

Five Dimensions of Narrative Comprehension: The Event-Indexing Model

Rolf A. Zwaan
The Florida State University

The successful comprehension of a narrative necessarily involves the construction of a coherent mental representation of the narrated events, states, and actions, a situation model (van Dijk & Kintsch, 1983). Trabasso and colleagues have provided clear evidence that causal relations play a central role in establishing these coherent representations (e.g., Trabasso & Magliano, 1996; Trabasso & Sperry, 1985; Trabasso & Suh, 1993; Trabasso & van den Broek, 1985). However, events in narratives are related on several other dimensions, for example, time, space, and protagonists. The work discussed in this chapter reports on progress being made toward an integrated account of how the multiple dimensions of situations play a role in narrative comprehension.

A large number of studies have provided empirical evidence that readers mentally represent the temporal, spatial, and protagonist-related connections between events (see Zwaan & Radvansky, 1998, for a review) in addition to the causal and motivational dimensions studied by Trabasso and others. These five situational dimensions (time, space, causation, motivation, and protagonist) are usually investigated in isolation by text-comprehension researchers. My colleagues and I have developed the *event-indexing model* (Zwaan, Langston, & Graesser, 1995; Zwaan & Radvansky, 1998) in an effort to examine how these dimensions operate in concert. I first provide an overview of the event-indexing model and then go into further detail about the assumptions behind it. Two general hypotheses about processing and representation can be derived from the model: the *processing-load hypothesis* and the *memory-organization hypothesis*. Empirical evidence bearing on these hypotheses is presented. I conclude with a discussion of future directions.

THE EVENT-INDEXING MODEL

According to the event-indexing model (Zwaan, Langston, & Graesser, 1995), comprehenders parse clauses of a text into events. During comprehension, they connect these events on five different situational dimensions: time, space, causation, motivation, and protagonist. If the event that is currently being processed overlaps with the events in working memory on a particular dimension, then a link between those events is established and stored in long-term memory. Overlap is determined based on whether or not two events share an index (i.e., a time, place, protagonist, cause, or goal).

Zwaan and Radvansky (1998) recently expanded this basic model. For example, they distinguished between: (a) the *current model,* the model currently under construction, that is, the model at t_n; (b) the *integrated model* of the situations at times t_1 through t_{n-1}; and (c) the *complete model* of the situations at times t_1 through t_x. The current model is constructed at time t_n while a person reads a particular clause or sentence, called c_n. The integrated model is the global model that was constructed by integrating, one at a time, the models that were constructed at times t_1 to t_{n-1} while the person reads clauses c_1 to c_{n-1}. Finally, the complete model is the model that is stored in long-term memory after all the textual input has been processed. Thus, the complete model is the same as the integrated model after all the input has been processed. Zwaan and Radvansky termed the process of incorporating the current model into the integrated model *updating.*

The processing-load hypothesis concerns the updating process. The more situational indices that are shared between the current model and the integrated model, the easier updating will be. The memory-organization hypothesis pertains to the structure of the integrated model and its final form, the complete model, in long-term memory.

There are four basic assumptions behind the event-indexing model:

1. Events are the central units of situation models (the event-centrality assumption).
2. Events can be linked on five dimensions: time, space, causation, motivation, protagonist (the five-dimensions assumption).

3. Events are either related or not related on a particular dimension (the dichotomy assumption).
4. All dimensions are equal (the dimensional equality assumption).

THE FOUR ASSUMPTIONS

The Event-Centrality Assumption

According to the event-centrality assumption, events are the core units of situation models. This is based on the fact that (a) events are the most dynamic aspects of situations and (b) every clause of a text obligatorily describes an event. It is important to address two questions: What are events? and How are they represented in memory? The answer to the first question requires a definition, whereas the answer to the second requires more empirical data.

WordNet®, the lexical database maintained by the Cognitive Science Laboratory at Princeton University, defines an event as "something that happens at a given place and time." This is a very reasonable and helpful definition. It states that events are fixated in time and space. This makes them different from, for example, people and objects, which can move through time and space. It also separates events from concepts, which are general notions, not bound by time or space. One drawback of the definition is that it tells us nothing about the grain size of events. For example, a raindrop falling on my hand would qualify as an event, but so would Woodstock and the Allied invasion in Normandy. Indeed, the Dutch poet Gerrit Kouwenaar titled one of his volumes *Landscapes and Other Events* (*Landschappen en andere gebeurtenissen*). And indeed, on a geological time scale, landscapes can be considered events. However, in a model of event comprehension, it is necessary to limit the scope of what can be called an event. As the examples show, some events are made up of multiple events, whereas others seem more like singular events. Research on the units of (visual) perception shows that people can zoom in and zoom out on events according to situational constraints, but that they are rather consistent at identifying events at each predetermined grain size (e.g., Newtson, 1973). The event-indexing model is mostly concerned with singular events, such as a falling raindrop, an explosion, or someone tripping over a cat, given that this is the most common type of events in narratives. However, the event-indexing model currently treats events of all types alike. This is likely to change as the model is developed further.

In its current form, the event-indexing model is most concerned with how links between events are formed in the comprehender's mind. Therefore, the model doesn't have a lot to say yet on how events themselves are represented in memory. However, a full-fledged model of situation-model construction obviously cannot simply treat events as empty nodes (see Johnson-Laird, Hermann, & Chaffin, 1984). It is plausible to assume that events have an abstract "kernel," somewhat akin to the semantic primitives in, for example, Schank's (1972) conceptual dependency model. These kernels can be augmented by specific event-related information. For example, "to eat" can be represented as "INGEST FOOD," a semantic primitive. Because eating a banana is very different from eating soup, and both are different from eating a steak, the kernel would be complemented with the relevant semantic and episodic information. Semantic information would include the knowledge that you first peel a banana before you eat it, that you use a spoon to eat soup and a fork and knife to eat a steak, as well as mental images of these foods. Episodic information would involve memories of eating these kinds of foods—for example, the odor, feel, and taste of a banana on a particular afternoon, burning your mouth when eating soup at your grandmother's house, and, perhaps, a guilty feeling when eating a steak last week.

Whether and how much episodic information would be incorporated depends, obviously, on the comprehender's background knowledge. Most of us would lack episodic information about performing heart surgery or reaching the summit of Mt. Everest. In these cases we may have to rely mostly on semantic knowledge. On the other hand, most of us know what it is like to give a lecture or send an e-mail message. So, when we read about someone sending an e-mail message, we may retrieve a mental image of our own computer screen and our own e-mail program from episodic memory and use it to augment the semantic primitive.

How much knowledge about an event is activated may depend not only on the reader's knowledge base but also on the way an event is conveyed verbally. In my laboratory, we have recently obtained evidence that certain linguistic structures may cue comprehenders about how much knowledge about an event they need to activate. Specifically, Zwaan and Stanfield (1999) found that, under certain conditions, comprehenders are more likely to activate mental representations of tools such as hammers when a related action, for example, pounding, is described as ongoing, as in the past progressive, than when it is described as instantaneous, as in the simple past tense. Thus, verb aspect appears to function as a cue to comprehenders regarding the amount of knowledge about an event that needs to be activated. When the event is described as ongoing, more knowledge is activated than when it is described as instantaneous.

The Five-Dimensions Assumption

We assume that comprehenders relate events to the evolving situation model on five dimensions: time, space, protagonist, causation, and motivation. In order to understand an event, we need to know at least the following: What happened When, Where, Why, and How and Who was involved in it? These questions map readily onto the event-indexing model. The *what* question refers to the event itself. *When, where, and why* refer to the temporal, spatial, causal-motivational, and protagonist dimensions, respectively. Finally, the *How* question was not captured in the original proposal (Zwaan, Langston, & Graesser, 1995), but would be incorporated as a label to the event in a version of the model that we are currently developing (Radvansky & Zwaan, 1998).

Why Are Causation and Motivation Separated? Most causal models in the text comprehension literature do not treat causation differently from motivation. That is to say, they usually do distinguish causation and intentionality theoretically, but not necessarily in their models. In the event-indexing model, causation and motivation are separate dimensions. One reason why it makes sense to have two separate dimensions is that it is possible for two events to be causally related or to be causally plus motivationally related. I can accidentally drop a hammer on your toe, which causes your toe to break. But I can also intentionally drop the hammer on your toe, for instance, when you and I are competing for a spot on the basketball team. In the first case, there is only a causal link (aside from the temporal and spatial links). In the second case, there is also a motivational link. Thus, the two events should be more strongly connected in the latter case than in the former, something that is most straightforwardly accommodated by a model that incorporates separate causal and motivational dimensions.

Why Is There Not a Separate Emotion-Based Dimension? Rather than questioning the separation of dimensions, one could question the selection of dimensions. For example, emotions seem a plausible candidate dimension. Research shows that readers make inferences about the emotions of protagonists and may do so even automatically (Gernsbacher, Hallada, & Robertson, 1998). Therefore, several people have suggested to us that the event-indexing model should include a separate emotion dimension. There are two reasons why we don't treat emotions as a separate dimension. First, many events described in narratives do not involve emotions, whereas all events involve time and space and many events involve causation and goals. Thus, connecting events based on emotions (and it is unclear how this should be done) would yield a rather sparsely interconnected network. Second, it would not be parsimonious to treat emotions as a separate dimension. Rather, emotions are deeply intertwined with goals (e.g., Stein & Levine, 1989). For example, one may get angry or sad when a plan fails or happy or proud when a plan succeeds. Therefore, emotions are treated as characteristics of protagonists that change as a result of goal success versus goal failure (Radvansky & Zwaan, 1998).

Although we do not think emotions should be regarded as a separate dimension, it is certainly possible that future research will dictate that more dimensions are to be added. However, in our view, it makes sense to add dimensions if only it turns out to be impossible or inelegant to explain findings with the current five dimensions.

The Dichotomy Assumption

Unlike the previous assumptions, the dichotomy assumption was prompted by methodological concerns, rather than theoretical ones. Thus far, we have treated situational dimensions as dichotomous: Either two events are related on a dimension or they are not. Although this simplifies matters empirically, the question is whether it is theoretically plausible. The answer to this question may be different for each dimension.

Time. In a dichotomous coding scheme, it does not matter if two events are 5 minutes apart or 5 years (all other things being equal). In Zwaan (1996), I obtained some initial evidence consistent with this scheme given that I did not find any statistically reliable differences between time intervals of 1 hour and 1 day, although there were some differences between the mean response times in these conditions. However, much more research is needed on the effects of the magnitude of time shifts on comprehension processes before the dichotomy issue regarding the temporal dimension can be addressed satisfactorily.

Space. In a detailed analog spatial model, readers would represent the distance between, for example, the protagonist and objects. In a more abstract representation, the reader would link objects that are close to the protagonist (e.g., in the same room) but not objects that are in different rooms, no matter how far or close those rooms are to the protagonist's current location. Recently, Rinck, Hähnel, Bower, and Glowalla (1997) found initial support for this position.

Causation. As mentioned earlier, Trabasso and his colleagues have developed sophisticated methods to analyze causation in narratives. For example, they distinguished enablements as a weak form of causation. In the event-indexing model, enablements were not considered causal relations. However, in unpublished data, we found that the fit of the model with data reflecting memory organization improved significantly when we augmented the causal analysis with enablements, but only when enablements received half the weight of the other causal relationships. This seems consistent with Trabasso's work and thus shows that the event-indexing model analysis could be refined. Moreover, van den Broek (1990) argued how the strength of causal relations may differ according to criteria of necessity and sufficiency. Thus, causal necessity and sufficiency may provide a good basis for converting the causal dimension from a dichotomous into a continuous one.

Goal. Goal-plan structures are conceived of as hierarchical representations (e.g., Graesser, 1981; Schank & Abelson, 1977), such that actions that are more relevant to the goal are higher in the hierarchy than less important actions. For example, when the goal is to buy a book, entering the bookstore is a relevant action but it is less important than paying for the book. Foss and Bower (1986) obtained initial evidence that the degree of priming between actions increased with the proximity of these actions within the goal hierarchy.

Protagonist. The protagonist dimension seems rather straightforward. Either a protagonist is the main agent in an event or he or she is not. However, the situation is not always that clear-cut. For example, some events have two or more agents, such as when people are kissing, hugging, arguing, fighting, debating, or slugging it out. In such a case, the event could be linked to other events that each of the protagonists was the main agent in. The question here is whether the link strength should be equal to that in the case of a single protagonist or whether it should be smaller. This question can be addressed empirically. An alternative would be to treat the cluster of participants as a new participant node, but this seems neither elegant nor plausible.

To summarize, it seems that the dichotomy assumption is likely to be an oversimplification and that different schemes might be needed for the different dimensions. However, a great deal more research is needed before a finer-grained system of determining the strengths of dimensional links can be developed. Until that time, the dichotomy assumption serves as a useful approximation.

The Dimensional Equality Assumption

The assumption that each dimension is weighted equally was, similar to the dichotomy assumption, made for methodological rather than theoretical reasons. However, as the various tests of the processing-load and the memory-organization hypothesis discussed in the next section show, the dimensions are not really equally weighted. We assume that readers can differentially weight the dimensions according to their reading goals. For example, if their goal is to form a spatial map, they may emphasize the spatial dimension more than under "normal" reading conditions (e.g., Zwaan & van Oostendorp, 1993). It may also be the case that some dimensions are constructed obligatorily. Questions involving the comparison among the situational dimensions are high on our empirical agenda.

HYPOTHESES AND EVIDENCE

The Processing-Load Hypothesis

The processing-load hypothesis holds that the integration of an event into the evolving situation model is facilitated to the extent that the event overlaps situationally with the current model. This assumption is not uncommon in the text comprehension literature. For example, Gernsbacher's (1990) structure-building framework also assumes that online comprehension is facilitated with situational overlap. However, the structure-building framework assumes that lack of situational overlap on any dimension will lead to an increase in comprehension time, because it will be a cue to the comprehender to build a new substructure. According to the event-indexing model, situational overlap is a matter of degree. Overlap can be on anywhere from zero to five dimensions. The greater the situational overlap, the greater the facilitation. Thus far, this hypothesis has been investigated in its full extent using only multiple-regression techniques. However, aspects of it have been investigated using factorial designs.

Zwaan, Magliano, and Graesser (1995) had college students read short stories one sentence at a time on a computer screen. The subjects pressed the space bar on the computer keyboard when they had read a sentence and the time between key presses represented the reading time for that sentence. The reading times were analyzed using multiple-regression techniques. In these analyses, a set of variables was used to predict reading times. Three of these variables represented dimensions of the situation model: time, space, and causation. In addition, variables that coded for surface level aspects of the sentences, such as the number of syllables and their serial position in the text, known to be robust predictors of reading times, were included as auxiliary measures. Thus, the effects of the three situational dimensions analyzed in this study were assessed after those of the auxiliary variables had been partialed out statistically. Zwaan, Magliano, and Graesser found reliable effects of temporal and causal discontinuities, but not spatial discontinuities, on reading times.

Recently, Zwaan, Radvansky, Hilliard, and Curiel (1998) provided an important extension of these results. They used the same paradigm as Zwaan, Magliano, and Graesser, (1995) but different materials. Furthermore, they analyzed all five situational dimensions rather than only three. In all three of their experiments, Zwaan, Radvansky, and their colleagues found reliable effects of temporal and causal discontinuities on reading times, thus replicating the results of Zwaan, Magliano, and Graesser. Moreover, in all three experiments, there were reliable effects of the goal and protagonist dimensions.

The pattern was less straightforward for the spatial dimension. As mentioned earlier, Zwaan, Magliano, and Graesser (1995) did not find effects of spatial discontinuities on reading times. Zwaan et al. (1998) obtained a similar finding in their first experiment. They reasoned that a lack of an effect for the spatial dimension might reflect one of two alternatives. Either readers do not monitor spatial continuity during narrative comprehension or the regression methodology is not sensitive enough to detect effects of spatial discontinuity.

To address these questions, Zwaan et al. (1998) used a paradigm introduced by Morrow, Greenspan, and Bower (1987) and further developed by

Rinck and Bower (1995). In the Morrow–Bower–Rinck paradigm, people first memorize the layout of a building before they read a set of narratives about events taking place in that building. Important to note, each text explicitly describes the movements of a protagonist from one room to another. Thus, the general logic of this paradigm is to first ensure that readers possess a detailed spatial model by having them learn a spatial layout of a building to criterion before reading the narratives about people moving about in that building.

The question of interest studied by Morrow, Bower, Rinck, and their colleagues is not so much whether or not readers *construct* spatially based situation models during comprehension. Rather, the question was whether and how readers *use* previously constructed mental maps during comprehension. However, Zwaan et al. (1998) were interested in whether spatial information has an influence on comprehension when it is not needed to respond to a memory probe or resolve an anaphoric reference. In their Experiments 2 and 3, they used the texts from Rinck and Bower (1995) and measured participants' sentence-reading times. In Experiment 2, the participants first memorized the layout of the building. In Experiment 3, they did not. Zwaan et al. found that spatial discontinuities were reliable predictors when people had first memorized the map (Experiment 2) but not when they had not (Experiment 3). The other four dimensions were reliable predictors in both experiments. Thus, whether or not people monitor spatial discontinuities appears to depend on the prior availability of a mental map.

Table 7.1 summarizes the results for the five experiments conducted in my lab that have employed multiple-regression analyses of reading times. The reading-time analyses have established reliable effects for time and causation in five out of five experiments, for goal and protagonist in three out of three experiments, and for space in only one out of five experiments.

TABLE 7.1

Effects of Situational Dimensions on Reading Times (Processing-Load Hypothesis)

Experiment	*Time*	*Space*	*Causation*	*Motivation*	*Protagonist*
ZMG Experiment 1	yes	no	yes	—	—
ZMG Experiment 2	yes	no	yes	—	—
ZRHC Experiment 1	yes	no	yes	yes	yes
ZRHC Experiment 2	yes	yes	yes	yes	yes
ZRHC Experiment 3	yes	no	yes	yes	yes

Note. ZMG = Zwaan, Magliano, and Graesser (1995). ZRHC = Zwaan, Radvansky, Hilliard, and Curiel (in press).

One advantage of multiple-regression analyses is that they allow the researcher to use naturalistic texts. Thus, it is possible to examine situation models in their "natural" complexity. On the other hand, it is sometimes difficult to isolate effects of specific variables using these methods. This can be achieved better in factorial designs. For instance, Zwaan (1996) examined, among other things, whether temporal discontinuities influence sentence-reading times. The materials consisted of a set of stories that came in three versions. Somewhere in the middle of the story, a sentence described a target event (e.g., "She walked onto the stage"). This sentence was followed by a sentence that started with a temporal marker ("a moment/hour/day later") and described an event that was causally unrelated to the previous event (e.g., "She collapsed"). Reading times for the hour and day sentences were significantly longer than those for the moment sentences (which were assumed to preserve temporal continuity). This pattern was markedly different from that in a control experiment in which the critical sentences were not embedded in stories. In this experiment, the reading times for the moment sentences were reliably *longer* than those for the moment and hour sentences. In other words, the reading times for the critical sentences obtained in the main experiments did not reflect at all their "base rate" reading times, thus eliminating the potential confound that the pattern of reading times in the main experiments was a result of differences in wording among the different versions of the critical sentences.

Many other researchers have examined the effects of causal, motivational, and protagonist-related discontinuities on reading times (see Zwaan & Radvansky, 1998, for a review). The evidence for the processing-load hypothesis is quite strong as far as time, causation, motivation, and protagonist go. However, the evidence regarding the spatial dimension is mixed. Current findings suggest that whether or not people monitor spatial information during narrative comprehension depends on whether (a) they have a previously constructed mental map available during comprehension or (b) they are explicitly instructed to focus on spatial information (e.g., Zwaan, 1993; Zwaan & van Oostendorp, 1993).

The Memory-Organization Hypothesis

According to the memory-organization hypothesis, the strengths of the connections between events in the comprehender's long-term memory are a function of the number of dimensions on which the events are connected. As with the processing-load hypothesis, this hypothesis has thus far been investigated in its full extent only using multiple-regression techniques. However, aspects of it have been investigated using factorial designs.

Zwaan, Langston, and Graesser (1995) developed a method to assess the strength of the various dimensions of the reader's situation model in long-term memory: the verb-clustering task. After reading a text, participants were presented with a set of verbs taken from that text. These verbs denote story events. The event-indexing model had been used in prior analyses of the texts to arrive at predicted verb-clustering scores on the five situational dimensions. The participants then clustered these verbs according to their memory for the text. The clustering scores for all possible verb pairs were subsequently analyzed using multiple-regression analyses. Among the predictor variables in these analyses were the five situational dimensions. The multiple-regression analyses provided estimates of the strength of the long-term memory links on each situational dimension. The results of several experiments that have used the clustering task are summarized in Table 7.2.

The results show some interesting trends. There are two experiments in which all five dimensions were significant predictors. Causation was a significant predictor in all seven experiments, motivation in six, protagonist and space in five, and time in four. Furthermore, in all experiments a majority of dimensions were reliable predictors. Moreover, in the experiments where two dimensions were nonsignificant, the experiment or experimental condition involved special populations, such as students with relatively low proficiency in the language of the narratives (Zwaan &

TABLE 7.2

Effects of Situational Dimensions on Clustering Scores (Memory-Organization Hypothesis)

Experiment	*Time*	*Space*	*Causation*	*Motivation*	*Protagonist*
ZLG Experiment 1	yes	yes	yes	yes	yes
Z&B (English texts)	yes	no	yes	yes	yes
Z&B (French texts)	yes	no	yes	no	yes
ZCD Experiment 1 (younger adults)	yes	yes	yes	yes	yes
ZCD Experiment 1 (older adults)	no	yes	yes	yes	no
ZCD Experiment 2 (younger adults)	no	yes	yes	yes	yes
ZCD Experiment 2 (older adults)	no	yes	yes	yes	no

Note. ZLG = Zwaan, Langston, and Graesser (1995). Z&B = Zwaan and Brown (1996). ZCD = Zwaan, Charness, and Dijkstra (1998).

Brown, 1996, French texts) or older adults (Zwaan, Charness, & Dijkstra, 1998), who were expected to construct weaker situation models, given their more limited processing resources compared to younger adults.

In memory-organization research, our factorial designs have thus far included only single dimensions. Specifically, we have used two different methods to assess the strengths of the connections between events in long-term memory. The first method is a primed-recognition task, which was originally developed by Ratcliff and McKoon (1978). The second is a memory-retrieval task that was originally developed by J. R. Anderson (1974) and later adapted by Radvansky and Zacks (1991) to study situation models.

In a primed-recognition task, participants read narratives and are presented afterwards with a list of statements for each text that they read. Their task is to indicate as quickly as possible whether or not those statements were from the text. The lists contain a prime-target pair, two statements from the text that are presented consecutively. The assumption is that subjects will access their long-term memory representations of the texts and corresponding situation models when making the recognition responses. Thus, when the prime item is read, subjects will activate the corresponding event node in their long-term memory representation. Activation will subsequently spread to events that are connected to the activated event. The stronger the connection between the prime event and the target event, the more priming the latter should receive and the faster the recognition responses to that item should be. The recognition responses to the target item should reflect the strength of the long-term memory connection between the prime and the target event.

Using this paradigm, Zwaan (1996, Experiment 3) showed that comprehenders connect events in their long-term memory based on temporal contiguity. Recognition responses to target statements (e.g., "She collapsed") presented after the prime event ("She walked onto the stage") were significantly faster when the events were temporally contiguous ("She walked onto the stage. A moment later she collapsed") than when they were not (e.g., "She walked onto the stage. An hour later she collapsed").

Recently, Hilliard and Zwaan (1998) used this paradigm to investigate long-term memory connections on the motivational dimension. Their study replicated and extended a study by Trabasso and Suh (1993; Suh & Trabasso, 1993). In this study, subjects were presented with one of two versions of a text. In both versions, a protagonist has a goal, such as buying a present for her mother's birthday. In the *goal-success* version, this goal is met halfway through the story and the protagonist sets up a new goal, for example, knitting a sweater for her mother. This goal is unrelated to the initial goal of wanting to buy a present. However, in the *goal-failure* condition, the goal is not met initially, for example, the protagonist does not have the money to buy a present, and the protagonist sets out to satisfy the initial goal of giving her mother a birthday present in a different way, for example, by knitting a sweater. Thus, in the goal-failure condition, wanting to give a birthday present and knitting a sweater are part of the same goal structure, but in the goal-success condition they are not. Trabasso and Suh's findings were consistent with this idea. When participants were comprehending the critical sentence about the protagonist engaging in a new action (e.g., knitting a sweater) the initial goal, giving a birthday present, was more activated in the goal-failure condition than in the goal-success condition. Thus, these findings suggest that a connection between the initial goal and the new action was more likely to be made during online comprehension in the goal-failure condition than in the goal-success condition, given that the two events were more likely to co-occur in working memory in the goal-failure condition.

Hilliard and Zwaan (1998) replicated and extended these effects using a primed recognition task (Experiment 1) and a primed-verification task (Experiment 2). In both experiments, participants read the 8 stories used by Trabasso and Suh, 8 new stories that were similar in structure to the Trabasso and Suh stories, as well as 20 filler stories. After the stories, the participants performed a primed recognition task as in Zwaan (1996). The primes were the goal sentences, for example, "giving a birthday present," and the targets were the new-action sentences (e.g., "knitting a sweater"). In both experiments, subjects who had received the instruction to read "normally" showed significant facilitation in the goal-failure condition compared to the goal-success condition.

Radvansky, Zwaan, Federico, and Franklin (1998) provided evidence for the memory-organization hypothesis for the temporal dimension using J. R. Anderson's (1974) memory-retrieval paradigm. In this paradigm, participants memorize a set of statements and are later on tested on their memory for these statements by way of a speeded recognition task. In their first experiment, Radvansky and colleagues presented their subjects with statements such as "The banker checked his watch when the camera flashed." All statements described the event of a protagonist performing an action (e.g., "checking his watch") plus an "anchoring event" (e.g., "the camera flashed") during which the action was performed. Among the list of studied items, there could be one, two, or three protagonist-actions paired with an event, that is, three different protagonists performing three different actions during the same event (e.g., the banker checking his watch, the plumber raising his glass, and the lawyer scratching his ear) or one action paired with one, two, or three different events (e.g., a camera flashing, an alarm sounding, a microwave beeping).

The idea behind this paradigm (see Radvansky & Zacks, 1991) is that memory retrieval is facilitated when events can be integrated into a single situation model. The memory-organization hypothesis predicts that events that take place at the same time should be more integratable than events that take place at different times, all other things being equal. Consistent with this hypothesis, Radvansky et al. (1998) found that response times increased reliably with the number of anchoring events that were linked to a single action whereas there was no reliable increase with the number of actions per anchoring event. Thus, when there were three sentences describing the lawyer scratching his ear at three different times (e.g., "when the telephone rang, when the alarm went off, when the camera flashed"), participants were experiencing retrieval interference, presumably because these three actions could not be integrated into a single situation model. On the other hand, when three actions were described that took place during one event (e.g., the lawyer scratching his ear, the banker checking his watch, and the sailor raising his glass when the camera flashed), participants integrated the three actions into a single situation model and experienced no appreciable retrieval interference. Thus, these results, like those of Zwaan (1996) and the clustering studies discussed earlier, show that people integrate events based on temporal overlap.

FUTURE DEVELOPMENTS

This overview of the assumptions behind, hypotheses generated from, and empirical evidence relevant to the event-indexing model shows that a great deal of work is still ahead of us, both theoretically and empirically.

In the theoretical domain, we need to learn more about how exactly the individual dimensions are represented and whether a uniform, dichotomous, representational format is the optimal solution. This is most likely not the case. Furthermore, we need to seriously consider the representation of the events themselves, rather than merely the nature of the relationships between them. We expect that the work in this area will make contact with research on autobiographical memory (e.g., S. J. Anderson & Conway, 1997), as well as with research on visual media (e.g., Zacks & Tversky, 1997). Researchers in these areas are also concerned with the mental representations of events. Although these events are not "acquired" through language, situation-model theory (e.g., van Dijk & Kintsch, 1983) predicts that there should be commonalties among how events are mentally represented, regardless of the mode of acquisition. Therefore, it might be fruitful for text-comprehension researchers to take note of these literatures, as others have observed as well (Taylor & Tversky, 1997). The implementation of the event-indexing model in a computer model would provide an important conceptual tool to flesh out the details of some of the assumptions made with respect to the model. In collaboration with Paul van den Broek, I am currently attempting to integrate the event-indexing model with the landscape model (van den Broek, Risden, Fletcher, & Thurlow, 1996).

In the empirical domain, there are many unresolved issues. For example, we need to learn more about the potential role of long-term working memory (Ericsson & Kintsch, 1995) in situation-model construction. Zwaan and Radvansky (1998) proposed that long-term working memory plays an essential role in situation-model construction. Long-term working memory holds promise as a theory of working memory that is more compatible with the demands of online discourse comprehension than traditional short-term working memory. Work is currently underway in my lab to examine this issue. Furthermore, as the previous discussion shows, we know very little about whether and how the situational dimensions interact. Given that regression-type approaches using naturalistic texts do not allow the researchers to control the frequencies with which links on particular combinations of dimensions occur, it is almost impossible to perform statistically powerful tests of potential interactions among situational dimensions. Factorial designs might provide a more promising avenue of research here. The only published study that examines combinations of situational dimensions using factorial designs that I am aware of is Taylor and Tversky (1997), who compared time, space, and protagonist. It will require a great deal of imagination, norming studies, and pilot work to come up with stimulus materials in which relationships are manipulated while relationships on other dimensions are kept constant.

ACKNOWLEDGMENTS

I thank Susan Goldman, Paul van den Broek, and Gabriel Radvansky for helpful comments on a previous version of this chapter.

REFERENCES

Anderson, J. R. (1974). Retrieval of propositional information from long-term memory. *Cognitive Psychology, 6,* 451–474.

Anderson, S. J., & Conway, M. A. (1997). Representations of autobiographical memory. In M. A. Conway (Ed.), *Cognitive models of memory* (pp. 217–246). Cambridge, MA: MIT Press.

Ericsson, K. A., & Kintsch, W. (1995). Long-term working memory. *Psychological Review, 12*(2), 211–245.

Foss, C. L., & Bower, G. H. (1986). Understanding actions in relation to goal. In N. E. Sharkey (Ed.), *Advances in cognitive science 1* (pp. 94–124). Chichester, England: Ellis Horwood.

Gernsbacher, M. A. (1990). *Language comprehension as structure building.* Hillsdale, NJ: Lawrence Erlbaum Associates.

Gernsbacher, M. A., Hallada, B. M., & Robertson, R. W. (1998). How automatically do readers infer fictional characters' emotional states? *Scientific Studies of Reading.*

Graesser, A. C. (1981). *Prose comprehension beyond the word.* New York/Berlin: Springer.

Hilliard, A. E., & Zwaan, R. A. (1998). *Reading tasks and global goal inferences.* Manuscript in preparation.

Johnson-Laird, P. N., Herrmann, D. J., & Chaffin, R. (1984). Only connections: A critique of semantic networks. *Psychological Bulletin, 96,* 292–315.

Morrow, D. G., Greenspan, S. L., & Bower, G. H. (1987). Accessibility and situation models in narrative comprehension. *Journal of Memory and Language, 26,* 165–187.

Newtson, D. (1973). Attribution and the unit of perception in ongoing behavior. *Journal of Personality and Social Psychology, 28,* 28–38.

Radvansky, G. A., & Zacks, R. T. (1991). Mental models and the fan effect. *Journal of Experimental Psychology: Learning, Memory, and Cognition, 17,* 940–953.

Radvansky, G. A., & Zwaan, R. A. (1998). *A new theory of situation models.* Manuscript in preparation.

Radvansky, G. A., Zwaan, R. A., Federico, T. & Franklin, N. (1998). Retrieval from temporally organized situation models. *Journal of Experimental Psychology: Learning, Memory, and Cognition, 24,* 1224–1237.

Ratcliff, R., & McKoon, G. (1978). Priming in item recognition: Evidence for the propositional structure of sentences. *Journal of Verbal Learning and Verbal Behavior, 17,* 403–417.

Rinck, M., & Bower, G. H. (1995). Anaphora resolution and the focus of attention in situation models. *Journal of Memory and Language, 34,* 110–131.

Rinck, M., Hähnel, A., Bower, G. H., & Glowalla, U. (1997). The metrics of spatial situation models. *Journal of Experimental Psychology: Learning, Memory, and Cognition, 23,* 622–637.

Schank, R. C. (1972). Conceptual dependency: A theory of natural language understanding. *Cognitive Psychology, 3,* 552–631.

Schank, R. C., & Abelson, R. (1977). *Scripts, plans, goals and understanding: An inquiry into human knowledge structures.* Hillsdale, NJ: Lawrence Erlbaum Associates.

Stein, N. L., & Levine, L. (1989). The causal organization of emotional knowledge: A developmental study. *Cognition and Emotion, 3,* 343–378.

Suh, S. Y., & Trabasso, T. (1993). Inferences during reading: Converging evidence from discourse analysis, talk-aloud protocols, and recognition priming, *Journal of Memory and Language, 32,* 279–300.

Taylor, H. M., & Tversky, B. (1997). Indexing events in memory: Evidence for index dominance. *Memory, 5,* 509–542.

Trabasso, T., & Magliano, J. P. (1996). Conscious understanding during comprehension. *Discourse Processes, 21,* 255–287.

Trabasso, T., & Sperry, L. L. (1985). Causal relatedness and importance of story events. *Journal of Memory and Language, 24,* 595–611.

Trabasso, T., & Suh, S. Y. (1993). Understanding text: Achieving explanatory coherence through on-line inferences and mental operations in working memory. *Discourse Processes, 16,* 3–34.

Trabasso, T., & van den Broek, P. W. (1985). Causal thinking and the representation of narrative events. *Journal of Memory and Language, 24,* 612–630.

van den Broek, P. (1990). Causal inferences and the comprehension of narrative texts. In A. C. Graesser & G. H. Bower (Eds.), *Inferences and text comprehension* (pp. 175–196). San Diego: Academic Press.

van den Broek, P., Risden, K., Fletcher, C. R., & Thurlow, R. (1996). A "Landscape" view of reading: fluctuating patterns of activation and the construction of a memory representation. In B. K. Britton & A. C. Graesser (Eds.), *Models of understanding text* (pp. 165–187). Mahwah, NJ: Lawrence Erlbaum Associates.

van Dijk, T. A., & Kintsch, W. (1983). *Strategies of discourse comprehension.* New York: Academic Press.

Zacks, J., & Tversky, B. (1997). The perception of event structure. *Abstracts of the Psychonomic Society, 2,* 61.

Zwaan, R. A. (1993). *Aspects of literary comprehension: A cognitive approach.* Amsterdam/Philadelphia: Benjamins.

Zwaan, R. A. (1996). Processing narrative time shifts. *Journal of Experimental Psychology: Learning, Memory, and Cognition, 22,* 1196–1207.

Zwaan, R. A., & Brown, C. M. (1996). The influence of language proficiency and comprehension skill on situational model construction. *Discourse Processes, 21,* 289–327.

Zwaan, R. A., Charness, N., & Dijkstra, K. (1998). Situation models, cognitive aging, and processing speed. Unpublished manuscript, Florida State University.

Zwaan, R. A., Langston, M. C., & Graesser, A. C. (1995). The construction of situation models in narrative comprehension: An event-indexing model. *Psychological Science, 6,* 292–297.

Zwaan, R. A., Magliano, J. P., & Graesser, A. C. (1995). Dimensions of situation model construction in narrative comprehension. *Journal of Experimental Psychology: Learning, Memory, and Cognition, 21,* 386–397.

Zwaan, R. A., & Radvansky, G. A. (1998). Situation models in language comprehension and memory. *Psychological Bulletin, 123,* 162–185.

Zwaan, R. A, Radvansky, G. A., Hilliard, A. E., & Curiel, J. M. (1998). Constructing multidimensional situation models during reading. *Scientific Studies of Reading, 2,* 199–220.

Zwaan, R. A., & Stanfield, R. A. (1999). *The resolution of situation models: aspectual and situational factors.* Manuscript in preparation.

Zwaan, R. A., & van Oostendorp, H. (1993). Do readers construct spatial representations in naturalistic story comprehension? *Discourse Processes, 16,* 125–143.

8

Goals as Generators of Activation in Narrative Understanding

Gordon H. Bower
Stanford University
Mike Rinck
Technical University, Dresden, Germany

Over the course of his highly productive career, Tom Trabasso has contributed greatly to advances in numerous areas of cognitive, developmental, and educational psychology. Indeed, the breadth and depth of his contributions is astonishing to contemplate and more than enough to have filled several research careers. The contributors to this volume honor Tom by writing about their research on discourse processing, an area in which Tom's research has had an appreciable and sustained impact. We have chosen to write about our research regarding "focus in mental models" of narrative discourse, and to relate our recent studies to the "causal path analysis" of narratives provided in very influential papers published by Tom and his associates in the mid-1980s.

One goal of research on language comprehension is to analyze the way in which readers (or listeners) construct and use internal mental models of the situation being described by the writer (or speaker). People interpret, react to, and remember texts or discourses according to the internal models they construct of them, and so these personal models are critical for explaining a person's behavior.

Research on mental models has advanced most rapidly in one specific area, namely, analysis of the way people comprehend simple narratives about human characters solving personal problems (Johnson-Laird, 1983; Sanford & Garrod, 1981; van Dijk & Kintsch, 1983; Zwaan & Radvansky,

1998). In reading a simple narrative, people construct an internal representation consisting of tokens of conceptual entities, including the settings, persons, objects, and instruments that play some role in the drama.

Readers use their model of the referential situation to interpret a text, to draw inferences, to connect together different parts of a text, and to update their knowledge of a changing situation. That is, they imagine the mental tokens (representing persons, entities) changing their relationships as the text describes the characters moving around in the settings, performing actions as they carry out plans to achieve their goals. Informally, we may identify the reader's mental model with the mental pictures that people report constructing and manipulating as they read a narrative. We suppose that readers construct and maintain their temporary mental model in a visuo-spatial working memory (which Baddeley, 1990, dubbed the "visuo-spatial sketchpad"). They use the incoming text as a guide to update their model of the current situation, to move characters from place to place and change the state of the story world.

The mental picture metaphor cannot be taken too literally because a mental model also contains readers' knowledge of invisible things, such as hidden objects (e.g., weapons, jewels) and actors' motives and emotions. In fact, a primary job of the model is keeping track of the goals of the characters and their fates—a topic elaborated later.

The mental model readers construct from a text is important psychologically because they remember their construction rather than the text itself. If the text is developing or adding to an ongoing news story, then its information will become assimilated into that developing knowledge. This assimilation is so automatic and powerful that readers may later be unable to identify the specific contributions of this text to their overall knowledge. They know the situation and unfolding events described but are likely to forget which sources contributed which parts of the knowledge.

In the following, we review several findings relevant to this mental-models approach. We first note the goal-directed nature of narrative episodes and how goals provide the glue that links narrative events into a coherent causal network. In this connection, we review the important research by Tom Trabasso and his colleagues on the causal analyses of narratives. We then examine one salient cause that links story events, namely, characters' motives (goals). We illustrate several properties of goals and actions that determine how easily readers will comprehend a character's actions. We then review our research on how readers' focus of attention within a mental map facilitates access to stored information regarding objects spatially near to the focus. We conjecture that this "distance from focus" effect arises from expectations that characters go to locations that are relevant to satisfying their goals. A final experiment is then described in support of this hypothesis, that goal relevance underlays the influence on memory access of distance-from-focus.

THREE MAJOR PARTS OF A NARRATIVE MODEL

The internal representation of a narrative consists of three major parts—the characters, the settings, and the events that take place in those settings. The model includes, first, descriptions of the cast of characters, their relationships, and possibly their occupations and personal traits. This information usually explains the goals and motives that drive the characters through the plot. The model includes, second, a sketchy mental map of the physical settings in which the actions occur. The setting may be named in only a vague manner ("One day while walking in the woods, ...") or named by calling up a standard, scripted setting (e.g., a cowboy-Western saloon, a war-combat zone, a Wall Street business meeting). The setting provides enabling (or constraining) conditions for the characters' actions. For example, cowboys in Western saloons of the 1850s could not use a telephone to call in reinforcements. Third, the model contains the temporal-causal sequence of episodes comprising the plot of the narrative. We begin by describing the nature of these episodes.

Stories Involve Characters Solving Problems

Simple narratives usually center around a main character who has a complicated problem to solve, and the story describes his or her actions in overcoming obstacles to the solution. Readers are presumed to act as "intuitive psychologists," trying to understand the actions of story characters in terms of lay theories. These theories suppose that people's actions are intentional and part of a deliberate plan aimed at achieving an outcome viewed as instrumental to some goal. Human goals tend to stem from a small set of desires and motives such as self-preservation, satisfaction of biological needs, achievement, entertainment, reciprocity, revenge, and so forth. Many motives and plans come in bundles associated with particular roles or occupations (e.g., a policeman, a doctor, a mother) and interpersonal themes and relationships (e.g., love, family, duty, enemies). Schank and Abelson (1977) and Wilensky (1978) wrote most thoroughly about the place of such themes and beliefs in readers' understanding of characters' actions in narratives.

A basic premise of Schank and Abelson (1977) is that readers of narratives are trying to explain characters' actions as part of a plan to accomplish a goal. Thus, for each action or event in a story, readers implicitly ask themselves: Is this action part of a plan that I know the actor is currently pursuing? If not, then is this the start of a new plan? If so, is it in service of an ongoing goal that I know the character has? If not an ongoing goal, is it a new goal that this character is likely to have? Is this new goal consistent with the roles or thematic relationships which I can attribute to this character? Schank and Abelson, and Wilensky (1978) in his PAM program (for "*P*lan *A*pplying *M*echanism"), analyzed the elaborate knowledge structures, beliefs, and inference rules people routinely use to explain and understand many plan-based phenomena. These phenomena include hierarchical subgoaling, goal substitution, goal interruption and suspension, goal resumption or abandonment, and so on. By using such explanations, readers build a network of causal connections among the events in the story—from some initiating events (e.g., a policeman stumbles across a robbery in progress) through the various goals, subgoals, and actions of the main character (e.g., he chases the robber), overcoming obstacles (e.g., the robber shoots at him), arriving at some final resolution (he captures and arrests the robber). Each goal is viewed as causing or enabling some actions that lead to outcomes that are evaluated relative to the protagonist's goal. The story may also exemplify some overall moral (e.g., "Crime doesn't pay").

The claim is that readers connect up events in narratives by finding or inferring these causal relationships, and the story is judged *coherent* to the extent that such relationships are found. Causal connections can be found either directly or by indirect inference from the immediately preceding text statements, or by remembering and reinstating information given much earlier in the text. Moreover, the ease of understanding a story varies with readers' ease of finding these causal connections.

This viewpoint has been adopted and defended by others (e.g., Bower, 1982; Graesser, Singer, & Trabasso, 1994; Sharkey & Bower, 1987). For example, the constructionist theory of Graesser et al. has as its main assumption that:

> [Readers] search after meaning. According to this principle, readers attempt to construct a meaning representation that addresses the reader's goals, that is coherent at both local and global levels, and that explains why actions, events, and states are mentioned in the story. (p. 371)

An important conjecture by Schank and Abelson (1977) was that the number of causal connections readers find in a story would play a central role in its psychological representation in the reader's interpretation and memory of it. They wrote:

> Thus connecting up the actual ... causalities is what makes sense of a text. A text is disconnected if a causal chain cannot be constructed to represent it. Furthermore, a theory of importance in text can be derived by causal chaining. Events or states that have multiple connections are likely to be highly significant with respect to the text. States or events that lead nowhere are likely to be forgotten. (p. 30).

This conjecture was one of several plausible ideas about what constitutes the "gist" of a text (for others, see Britton & Black, 1985; Meyer, 1975). The issue was effectively resolved by important research by Tom Trabasso and his associates (Trabasso & Sperry, 1985; Trabasso & van den Broek, 1985; Trabasso, van den Broek, & Suh, 1989). Whereas Schank and Abelson (1977) illustrated causal analysis for only "toy" vignettes of two or three lines, the Trabasso and Sperry article was significant in developing an operational system and criteria for coding longer, realistic narratives (actual folk tales) into causal networks. In their initial work, they applied the system to six folk tales (from a study by Brown & Smiley, 1977) so that each clause in each story was characterized as being on (or not on) the main causal chain that coursed through the narrative. Each clause was also scored for the number of causal connections it had—that is, how many other story actions or states it caused or enabled, or how many caused or enabled it. An example of the causal network for one of these stories ("The Father, His Son, and Their Donkey") is shown in Fig. 8.1. The numbered nodes refer to successive clauses of the story, and the arrows reflect direct causal connections or enablements between the numbered clauses. In such networks, it is obvious that some clauses have very few connections (e.g., clauses numbered 7, 8, 49), whereas others have very many (e.g., clauses 2, 25, 62).

Brown and Smiley (1977) collected "importance" ratings from their adult participants for each clause in each story. Importance was defined somewhat vaguely as the centrality of the given clause in contributing to the overall meaning of the story. Trabasso and Sperry (1985) wanted to predict these importance ratings from their causal network analyses of the six folktales. Figure 8.2 from Trabasso and Sperry shows the average importance rating for clauses as a function of the number of causal connec-

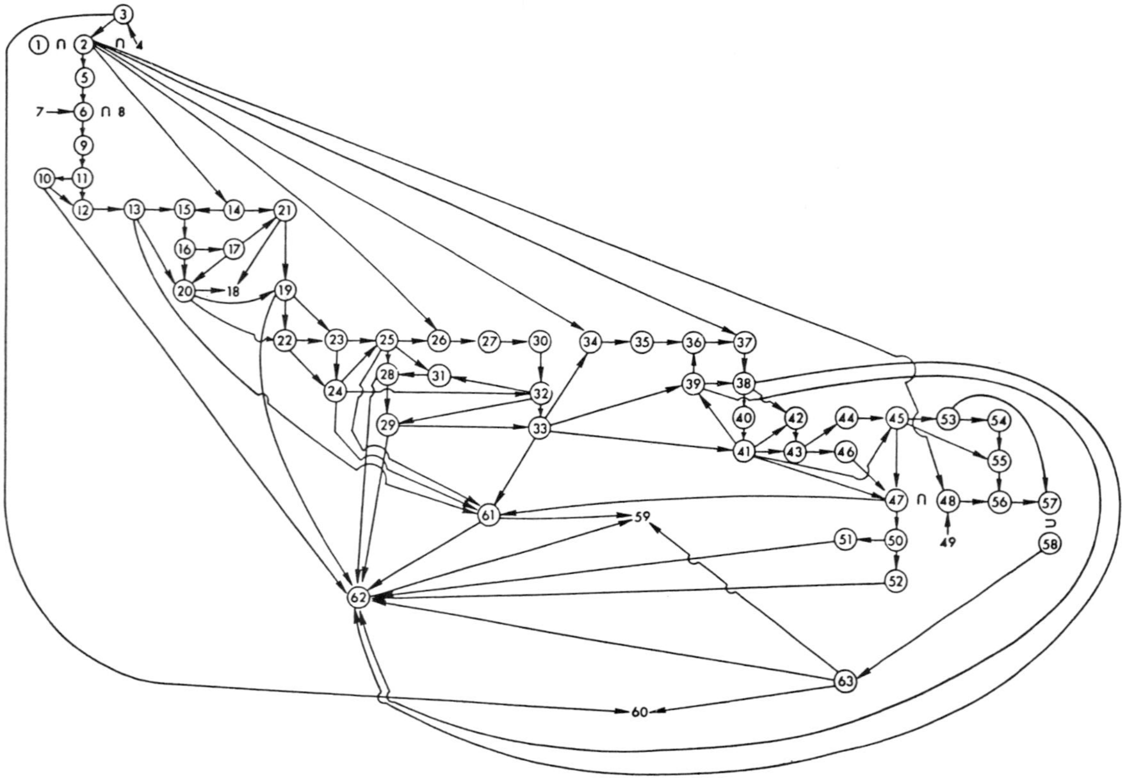

FIG. 8.1. The network linking clauses (numbered nodes) via cause or enablement linkages in a causal analysis of the folk tale “The Father, His Son and Their Donkey”. From Trabasso and Sperry (1985). Copyright 1985 by Academic Press Inc. Reprinted by permission.

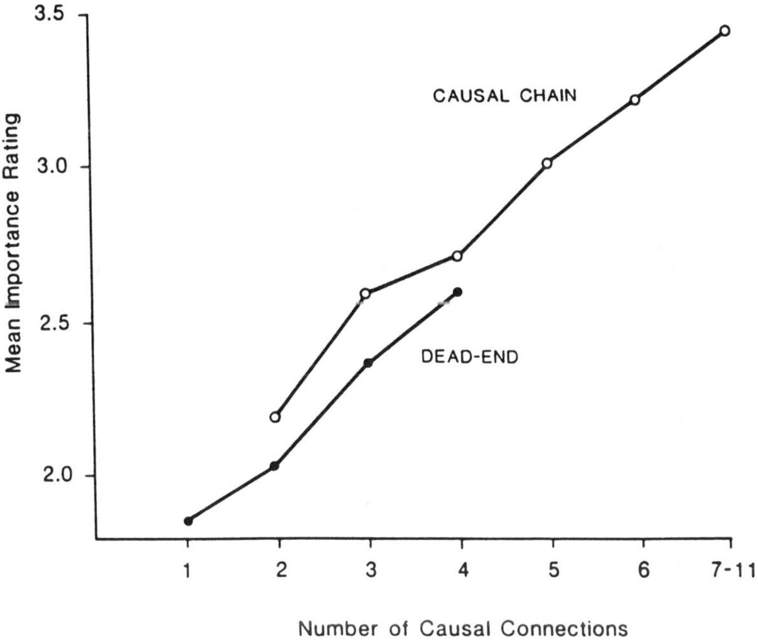

FIG. 8.2. Average importance judgment values for dead-end and causal chain statements that vary in the number of direct connections. From Trabasso and Sperry (1985). Copyright 1985 by Academic Press Inc. Reprinted by permission.

tions for the clause. For events on the main causal chain as well as for those on “dcad-cnd” sidc branchcs, a strong lincar rclation appcarcd bctwccn thc importance rating of a clause and its number of causal connections. Across the six folk tales, the number of connections accounted for a large portion of the variance in importance ratings of the clauses. Moreover, in other work, Trabasso and his associates showed that the number of causal connections accounted for which clauses were likely to be recalled and also included in a brief summary of the story.

These results represent a very significant advance in our understanding of story memory. Psychologists have known for a long time (certainly since studies by Bartlett, 1932) that people tend to recall mainly the gist or important ideas of a text. However, that generalization is incomplete until we can specify what are the important ideas in a text. Trabasso's causal analysis provides us with a fairly mechanical procedure for deciding what are the important ideas in a story; this can be done by simply analyzing the causal and semantic structure of the narrative itself. This analysis represents a remarkable advance over the earlier days of Bartlett when psychologists had no procedure for specifying the gist of a story.

Effects of Goal-Action Relatedness

The power of causal linkages in determining story representation leads to an interest in analyzing various kinds of causes—a topic to which Trabasso and his colleagues have made important contributions (Trabasso & Nickels, 1992; Trabasso et al., 1989). Causes can be either necessary but not sufficient ("enabling states") or directly sufficient in the circumstances, or both. Causes can be either physical ones (gravity, earthquakes, tornadoes) or psychological ones. The results can be either a state change of some physical object, or a mental change of an animate agent. The mental change can either enable or directly cause (or "be a reason for") an action.

We believe that the most important causes of actions in stories are characters' goals. In general, goals generate intended plans, and actions occur when plans are activated. But there are often multiple plans that will satisfy a given goal, and people have appreciable common-sense knowledge about what kinds of plans work for what kind of goals. (Much of the Schank & Abelson, 1977, book is devoted to describing this knowledge.)

How rapidly an action is understood depends on how expected and familiar it is as a method for achieving a known goal of the actor. In earlier work carried out with Carolyn Foss (Bower, 1982; Foss & Bower, 1986), this intuitive notion of goal-to-action relatedness was translated into the notion of psychological "distance" or the number of steps between a goal and an action in a subgoal (or plan) hierarchy. The idea was that the more subgoals and enablements readers needed to infer in order to fill in the connection between the actor's goal and his action, the longer readers would take to understand that action. Thus, from the goal, "John was hungry," it is a close, familiar step to "He ate a pizza." On the other hand, a far longer inference is required to relate that hunger motive to a remote action such as "He took out the yellow pages"—presumably to look up the address or phone number of a restaurant.

This relatedness notion was tested in an experiment in which college students read short, four-line vignettes in which a character was introduced, was assigned one of two goals, and then performed an action (Foss & Bower, 1986). Participants paced their line-by-line reading by pressing a button to present successive sentences of the vignettes to themselves. Participants' reading time of the action statement presumably indicates the time they required to comprehend the statement and integrate it into their interpretation of the text. The action statements were fixed, but across vignettes different participants read a preceding goal statement that was either psychologically "near" to (closely related) or "far" from the target action. Two example vignettes are shown next; a given participant read either the Near or the Far goal statement:

> (Opening) John was unemployed and desperate.
> (Near) He decided to rob a bank.
> (Far) He decided to steal some money.
> (Action) He checked out the security guards at the Wells Fargo Branch.
> (Ending) He tried to round up some accomplices.

> (Opening) Marge was concerned about the environment.
> (Near) She decided to participate in an anti-nuke protest rally.
> (Far) She wanted to stop construction of a nuclear power plant.
> (Action) She painted a protest sign.
> (Ending) Marge believed atomic power plants are harmful.

Each vignette was followed by two easily answered, true–false questions to ensure comprehension. In developing the materials, Foss and Bower tried to select Near goals that would be nested within the Far goal-to-action chain of subgoals. For example, robbing a bank is a more specific subgoal of the top goal of stealing money; participating in an antinuke protest rally is a more specific subgoal of stopping construction of a nuclear power plant.

Fourteen Stanford University students participated in this experiment reading 48 vignettes, with half the participants reading the Near and half the Far goal version of each vignette. Their action reading times confirmed the predictions: Participants took an average of 2.10 seconds to read the target action following the Far goal, but only 1.78 seconds to read it following the Near goal. Therefore, as predicted, participants needed more time to integrate a target (action) sentence into an ongoing goal-plan the greater was its inferential distance from the actor's stated goal. The result supports the general notion that people comprehend actions by connecting them to the actor's goals.

The Number of Goals

Schank and Abelson (1977) also proposed that story understanders set up a goal list for each character, and monitor this goal list as they progress through the narrative. Readers monitor what goals a character probably has by virtue of his or her roles, occupation, and thematic relationships to other characters (e.g., relatives, loved ones). Readers add new goals to the list as circumstances create them; they also delete old goals from a character's goal stack as they are achieved or abandoned; still other goals may simply be left unresolved by the story's end. In this way readers construct and continuously update what Schank and Abelson called a "goal-fate" chart for each character. The evaluation and significance of story events is largely reflected in the goal-fate graph of the main characters.

If readers are indeed explaining and evaluating actions in light of a goal list for an actor, it is reasonable to suppose that the time readers require to check the fit of an action to an actor's goals would increase with the number of goals the story has made active for that character at the time. The hypothesis is that when a target action is encountered in the text, the reader must "scan the goal stack" in memory for that character to see if he has at least one goal that would explain that action. If no such goal can be found in that character's stack, then the action will be inexplicable and the story possibly incoherent at this point. The prediction here is very much like that seen in the list-scanning task of Sternberg (1966), except participants are judging whether successive goals could explain the action probe.

To test this implication, Sharkey and Bower (1987; cited in Bower, 1982) presented readers with a goal-action decision task. On each of a series of trials, participants were presented with one, three, or five independent goals being attributed to a novel character. The goals were presented as single statements; an example of a three-goal series might be: "Heather *wanted* to overcome her shyness; she *wanted* to live extravagantly; she *wanted* to have some time on her own." Each goal ("wanted") statement was presented for 5 seconds. At the end of the one, three, or five goal statements, the computer beeped and then presented a test action statement. This named the actor and an action in the format "And so (Heather) (took classes in social skills)." The participants' task was to decide as quickly as possible whether that action satisfied one of the goals they had just read for this character. The decisions were easy in that a given action was either clearly consistent with one of the goals (e.g., classes in social skills help overcome shyness) or was inconsistent and completely independent of them (e.g., "So Heather lit a match" would not obviously satisfy any of the goals). Participants judged a large number of such trials, half of which had "Yes, True" answers and half of which had "No, False" answers.

The average decision times for both true and false answers are shown in Fig. 8.3. The decision times increase with the number of goals to be searched

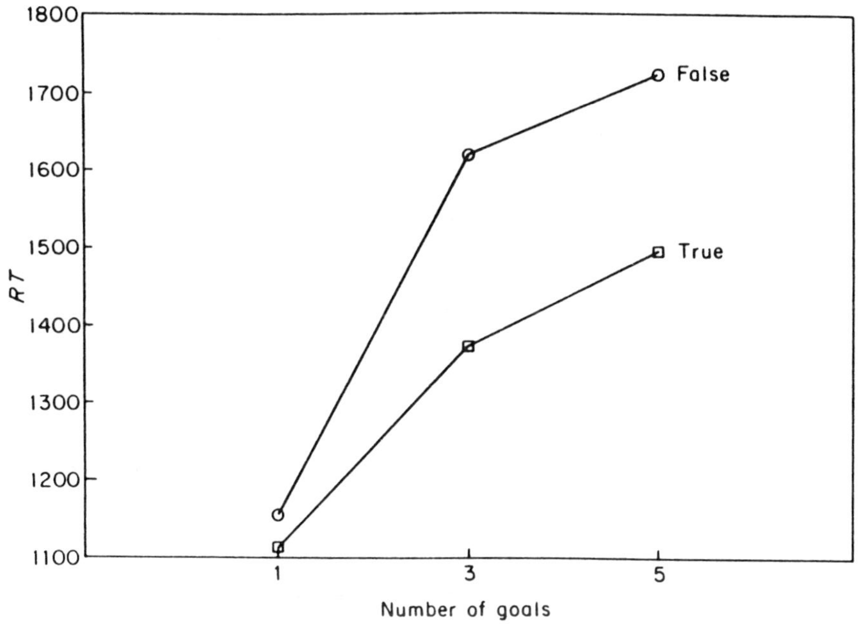

FIG. 8.3. Mean time to judge that an action was (true) or was not (false) consonant with some one of one, three, or five independent goals ascribed to a character. From Sharkey and Bower (1987). Copyright 1987 by John Wiley & sons, Limited. Reprinted by permission.

through for that character. It is as though the participant conducts a self-terminating search through that character's goal stack, checking one after another goal to see whether it would explain the action. The search terminates when an explanatory goal is found (which explains the faster "true" decisions); otherwise, the search proceeds to the end of the goal list at which time a "false" judgment is rendered. The effect here is akin to Sternberg's list-scanning time or the "fan effect" (Anderson, 1976), with the proviso that some reasoning or stored knowledge is needed to relate a goal to an action.

In a later experiment, Sharkey and Bower (1987) varied whether the three goals assigned to a character were independent or whether they could all be satisfied by a single action (were "related"). As an example, the three goals might be "Brian *wanted* to be of service to the community, *wanted* to drive a big shiny truck, and *wanted* to wear a uniform." These would all be satisfied by the action "And so Brian became a fireman." With these related goals, the action will satisfy any (and all) of the three goals. In such cases, decisions were expected to be very rapid. And results showed that indeed they were: In the experiment, decision times averaged 1,230 ms for a single goal, 1,475 ms for three independent goals, but only 1,240 ms for three related goals for which the action satisfied any (and all) of the three. Decision times for three related goals equaled that for one goal and was faster than that for three independent goals.

Clearly, the slower decision time for three goals was not due to a greater memory load for three compared to one goal. Rather, it appeared due to participants having to search through the stack of three goals when they were independent but not when they were related (for which any goal produces a *yes* answer). This analysis was further bolstered by testing actions that satisfied just one of the three related goals (e.g., the action "And so Brian became a social worker" satisfies the service goal but not the other two goals). In this case, we found that the decision times increased to equal those observed for three independent goal-action units.

The results from this series of experiments thus consistently support the general view that readers track characters' goals by keeping active in memory an associated list of goals, and that they try to understand a character's actions by searching through his or her goal stack. The more goals to be searched and rejected, the slower the comprehension time; actions that satisfy several goals at once are understood more readily as a consequence.

SPATIAL SITUATION MODELS

The second aspect of models of stories consists of the representation of spatial situations. The spatial model includes a more or less sketchy mental map of the places, landmarks, and objects as they are laid out in space and the locations of the characters as they move about. In our research, we have examined the properties of the spatial model as it is updated by readers as they comprehend story events. Specifically, we have studied how readers update their mental model as they follow the actions of the characters from one place to another in pursuit of their goals. Our approach assumes that readers construct in imagination a sort of theater stage or "dollhouse" with landmarks and rooms filled with expected objects, plus any special objects mentioned in the text.

Focus in Spatial Models

Storytellers relate their account from a specific perspective, usually that of a main character (protagonist) whose goals are accorded priority and importance. In describing the sequence of story events, the storyteller sweeps the "here-and-now" of the narration point across an ordered progression of episodes. The protagonist's movements in space may be thought of as causing in the reader a shifting "spot of light" or focus of attention that moves over corresponding parts of the mental model. The objects in this internal focus of attention are momentarily highlighted or foregrounded. Our notion of "focus" is much like the concept of "foregrounding" proposed by the linguist, Wally Chafe (1972, 1994). We explore some implications of this focus hypothesis for processing; specifically, we believe that concepts and entities in the focus of attention should be more active in memory and thus more readily retrieved.

We have used a spatial priming methodology for studying the influence of this shifting focus of attention. Objects currently in the focus of attention should be more accessible than other known objects in the mental map. This increased accessibility may be conceived as reflecting subconscious activation of the representations of objects in or near the focus of attention in the mental map. Moreover, objects that were recently in the focus should have a decaying level of activation, declining in accessibility as time passes since they were last in focus.

Starting with research by Morrow, Greenspan, and Bower (1987), a program of studies has provided considerable evidence in favor of this hypothesis; namely, items in the current focus of a mental model are more activated as revealed by priming. We use an experiment by Morrow, Bower, and Greenspan (1989) to illustrate the effect. In this experiment, college student participants first memorized a map of a research-laboratory building (see Fig. 8.4) of 10 rooms each containing four named objects. Participants then sat before a computer screen and presented themselves with a series of 20-line stories in which a protagonist was introduced and given some goal (e.g., to search the building for some lost notes, or an intruder, or to clean the rooms) that required him or her to walk around from room to room in this building. Our hypothesis caused us to be especially interested in the "movement" sentences, such as "Wilbur walked from the conference room into the library" (see Fig. 8.4). The place the protagonist just left is called the *source* room, the place he passes through is called the *path* room, and the place where he ends up after the motion event is called the *goal* room. Any other room in the building is called an *other* room.

We expected the reader's mental representation of the goal room, on which the focus is now centered, to be most active; the path and source rooms were expected to be less active than the goal room but more active than other rooms of the building more distant from the motion event.

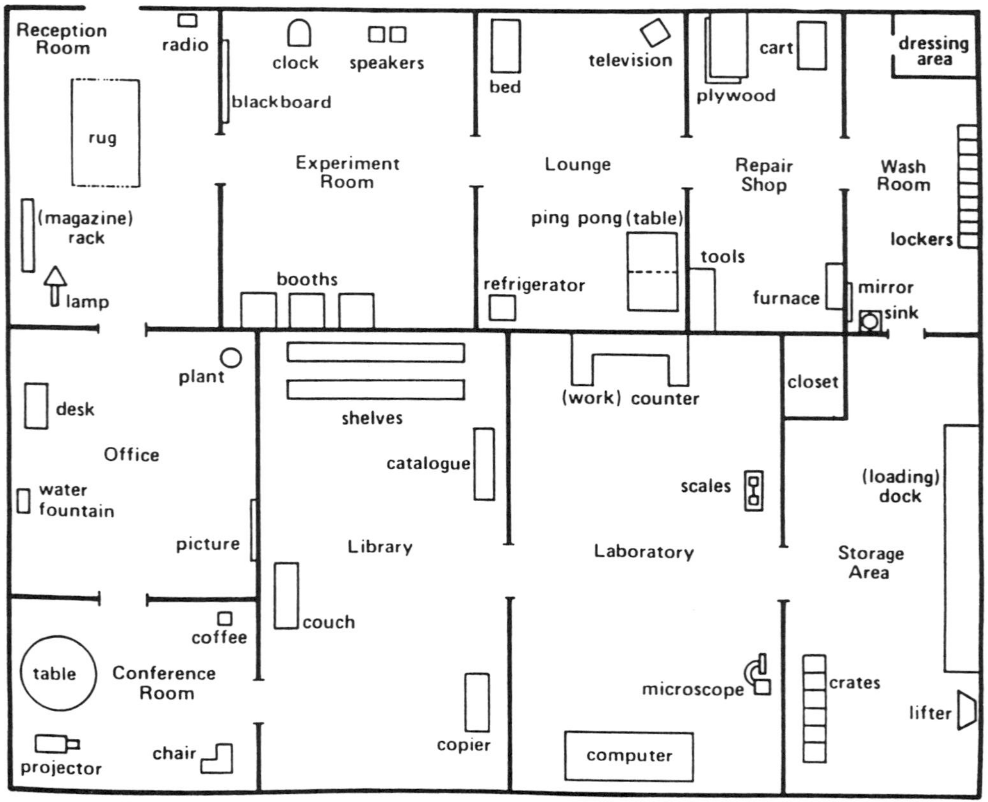

FIG. 8.4. An example of a 10-room research building, each room with four named objects, memorized by subjects. From Morrow, Bower, and Greenspan (1989). Copyright 1989 by Academic Press Inc. Reprinted by permission.

To test these predictions, we examined participants' times to answer simple questions about the locations of objects in the map they had memorized earlier. Periodically, the participants' line-by-line reading was interrupted by a pair of test words (naming objects in the building) rather than the next line of text. Participants were instructed to decide whether the two objects were in the same room (with each other) or in different rooms, and so indicate by pressing one of two keys marked *same* or *different*. Such probe tests occurred two to four times per story; they followed most of the motion sentences, but also occurred at other places in the text to prevent participants from giving special attention to the motion sentences. Several of the probes tested for the current location of the protagonist (e.g., "*Wilbur, Copier*") to ensure that participants tracked his current location. Half the probes were same-room and half different-room. At the end of each story, participants answered two easy yes–no questions about it to ensure that they had read for comprehension. Participants who made more than 10% errors on these comprehension questions were excluded from the analysis.

The average probe reaction times for same-room probes in this experiment are shown in Figure 5. This shows an orderly gradient: Objects in the goal room are most accessible; those in the path room are next most accessible; those in the source room are next; and objects in other rooms distant from the current focus are least accessible. The fact that objects in the path room are relatively more activated suggests that readers are leaving tracks or "a trail of activation" along a path as they imagine the character moving over those intermediate places. This effect arises despite the fact that the path room was not explicitly mentioned in the motion sentence just before the probe test.

In later experiments, Morrow et al. (1989) asked whether the reader's attentional focus is determined by the protagonist's physical location or by his "mental location," the place the protagonist is thinking about. Having located the protagonist in some room (call it A), the critical sentences were of the form: The protagonist "was thinking about (doing some activity) in room B," such as painting the walls there or repairing the furniture there. Probe questions would then occur testing the accessibility of objects in the protagonist's physical location, mental location, or some other room of the building. The results of that experiment showed shorter answer times for questions about objects in the character's mental location than for questions about his physical location. Thus, it appears that readers take the main character's perspective; they follow the character's thoughts, activating the same concepts that the character is thinking about.

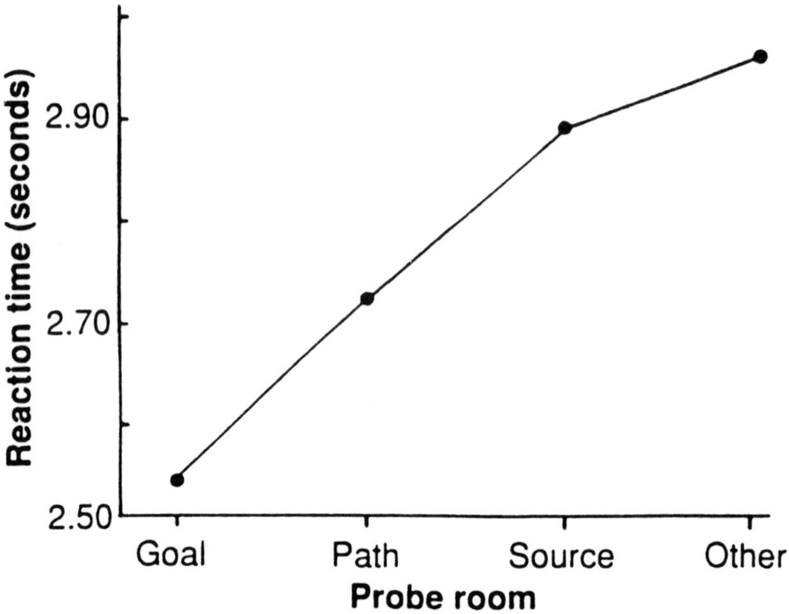

FIG. 8.5. Same-room decision times for pairs of objects in the goal, path, source, or some "other" room in the building. From Bower and Morrow (1990). Copyright 1990 by American Association for the Advancement of Science. Reprinted by permission.

Spatial Gradient in Anaphor Resolution

The method for measuring priming in the foregoing experiments was "two-object" probes, asking "Are Object 1 and Object 2 in the same room?." The disadvantage of this method is that it interrupts the normal flow of reading and emphasizes for readers the locations of objects and the protagonist. A less intrusive measure was introduced by Rinck and Bower (1995), who simply recorded participants' time to read critical sentences that contained an anaphoric reference to some object in the building. The anaphor was the mention of the object in a definite noun phrase, as in "*the* copier" or even "*the* copier in *the* library." In order to refer to objects in rooms differing from the location of the protagonist, we adopted the "thinking about" format used in the previous experiment. That is, after having moved the protagonist from source room A through path room B into goal room C, the story would then include a statement such as "(The protagonist) just then remembered that he had to clean up [the *object*] [in the *location*]," as in "Wilbur remembered he had to clean up *the shelves* in *the library*." The *object* was chosen to be at varying distances (number of intervening rooms) away from the protagonist's location, and was always a unique object in the map memorized earlier. The critical sentence always used a conceptualizing verb such as *remembered, thought,* or *decided.* In different conditions, the phrase mentioning the location room was added or not. The crucial behavioral measure was simply how long participants took to read and comprehend such critical anaphor sentences, as they paced themselves line-by-line through the story by pressing the "mouse button."

The results across a number of experiments have been exceedingly orderly and regular. Figure 8.6 shows the results from one experiment: Reading times for critical anaphor sentences are separated according to whether they did or did not contain the location-room phrase. The total reading time for differing types of sentences has been adjusted by dividing total time by the length (in syllables) of the sentence, so that time per syllable is plotted in Fig. 8.6.

The results show a strong effect of location (room) of the referent as well as facilitation due to mentioning the room containing the anaphoric object. The room location (goal, path, source, other) is just a proxy for the distance

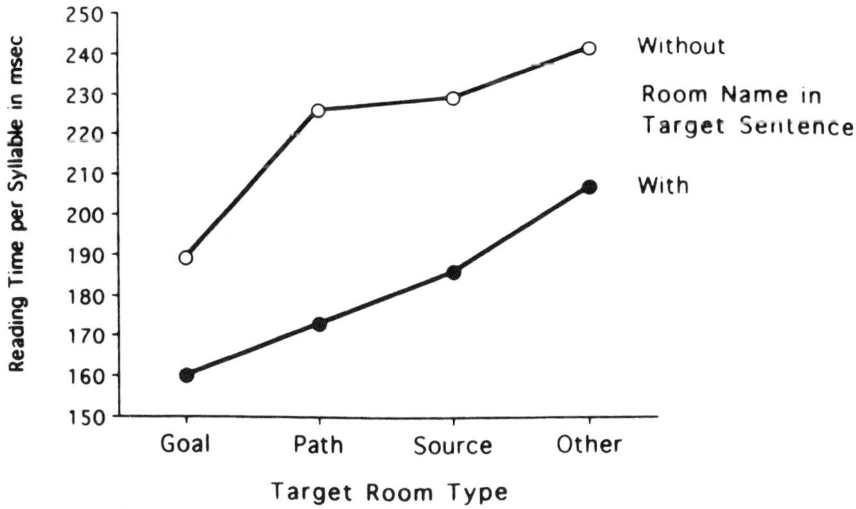

FIG. 8.6. Mean reading times per syllable (in milliseconds) of target sentences with and without the target room name, categorized by target room type. From Rinck and Bower (1995). Copyright 1995 by Academic Press Inc. Reprinted by permission.

between the locations of the current focus (on the protagonist) and the object referred to by the target sentence. The speed-up due to mention of the referent's room is understandable if the room name is viewed as a second cue helping retrieve from memory the representation of the anaphoric object. In later research, we have exploited this anaphor technique to investigate several issues regarding access to features of situational models. However, space limitations preclude us from reviewing that research.

Goal Activation of Model Objects

In the foregoing experiments, we have shown differential activation of objects depending on whether they were foregrounded as something the protagonist was thinking about. Most assuredly, one way to get readers to appreciate what is on the protagonist's mind is to attribute an active goal (wish, desire, plan) to him or her. Thus, if the story states that the protagonist was hungry, we would expect that items associated with eating would be assigned priority in the protagonist's mind and consequently in the reader's model of the protagonist. In particular, memorized representations of food items in the building should be activated and ready to be referred to. Similarly, if the protagonist needed to make a call, telephones should be primed; if he needed to sweep up, brooms should be primed; if he needed to check his slides, the projector should be primed; and so on.

An experiment to test this conjecture was carried out at Stanford University with the able assistance of Erin Graves. We describe this experiment in some detail, because it has not been previously reported. The experiment orthogonally varied the distance from the protagonist to the probe object and the relevance of the probe object to a recently mentioned goal of the protagonist. The experiment followed a simple 2 × 2 design with spatial distance (near vs. far objects) and goal relevance (relevant vs. irrelevant objects) varied as within-subject factors. The 40 participants first memorized the layout of a 10-room research building (similar to Fig. 8.4) with two named objects in each room. They then read 15 unrelated stories (3 practice, 12 experimental stories) with the stories presented sentence-by-sentence self-paced on a computer screen and sentence reading times were recorded.

The 12 crucial narratives contained 20 experimental "blocks" of critical sentences. Each block started with a three-room motion sentence, moving the protagonist from a source room through a path room to a goal (or current location) room, such as "Then he walked from the repair shop into the experiment room." Next, a pair of brief sentences attributed some goal or desire to the protagonist. This goal would require use of some object that was either in the location room (near) or in the source room (far). As one example, the pair of goal statements might be: "Looking at his list, he noticed that he had wanted to make copies of his presentation hand-out for the directors. He decided to make the copies immediately before he forgot." In this example, the goal object (*the Xerox machine*) is in the current location of the protagonist. Alternatively, the pair of goal statements might require use of an object in the source (far) room. Assuming the protagonist is now located in the experiment room, an example of a distant goal prime would be: "He remembered that he wanted to practice his presentation by videotaping the talk he had to give to the directors. He decided to go and start setting that up immediately before he forgot." These sentences should activate the *camcorder* that is located in the repair shop, which was the source room which the protagonist had recently left. Notice that the goal sentences never mentioned the goal object (*Xerox machine* or *camcorder*) explicitly, but rather only implied it, as we were interested in goal-object activation by relevance, not in simple name repetition.

Immediately after the goal sentences, the accessibility of objects was measured by "object-in-room" probes, asking participants to indicate whether they had learned earlier that the named object was in the named room. Participants pressed a yes or no key on the keyboard as quickly as they could to indicate their decision. For the 20 critical probes, the correct answer was always yes; many filler probes were used at other places in the stories to equate the number of yes and no answers. As noted, in the critical probes, the object could be either near (location room) or far (source room), and relevant or irrelevant to the preceding goal. To illustrate, if the goal were to videotape one's lecture, then *camcorder* would be relevant and far, whereas *Xerox machine* would be near and irrelevant. But if the goal were to copy some notes, then *camcorder* would be irrelevant and far, whereas *Xerox machine* would be near and relevant. Each participant judged five test probes in each of the four (2 × 2) conditions. To counterbalance experimental conditions and materials, each object was tested across participants equally often in each of the four conditions. Errors were well below 10% for all probe conditions.

The average results for these four probe decision times are shown in Fig. 8.7. Here we see strong effects of both factors on decision times: There is a distance effect, with the location of objects in the near ("location") room being verified faster than those in the far ("source") room ($F(1,39) = 5.84$, $p < .05$); there is also a strong effect of goal relevance, with objects relevant to a current goal being primed more strongly ($F(1,39) = 5.96, p < .05$); and the two factors are additive, with no interaction between them in this experiment ($F(1,39) < 1$). Figure 8.7 displays the main result we were seeking, and it reveals that both factors are important in activating objects in memory representation.

A second finding in this experiment concerns the time participants took to read and comprehend the sentence that immediately followed the probe test. Each of these "following" sentences provided a smooth continuation of the story whichever of the goal statements had been introduced earlier. This following sentence was usually of the form, "After he had finished that task, he [returned and] made sure that the experimenters would be busy conducting studies tomorrow so the directors would see how industrious they were." For different goal participants, of course, "that task" was referring to either Xeroxing or videotaping; if the goal activity took the protagonist back to the source room, then the "returned" verb was inserted into the following sentence as noted earlier. Filler words were added to the following sentence so that it had approximately the same number of syllables regardless of the location of the preceding goal-activity.

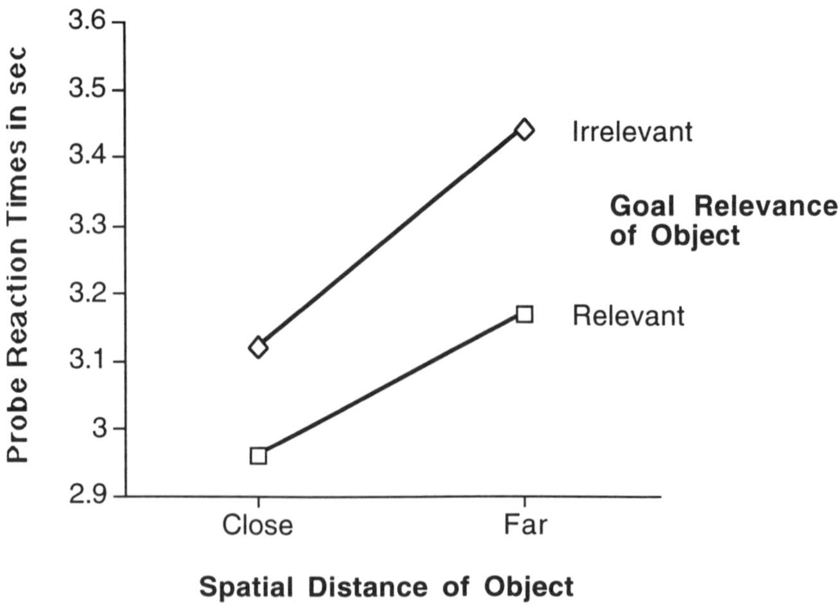

FIG. 8.7. Time to answer "Object X in Room Y?" questions depending on whether object X is in the Location or Source room and whether it is relevant or irrelevant to the protagonist's current goal. From an experiment by Rinck, Bower, and Graves.

The average reading times for these "following" sentences are shown in Fig. 8.8 grouped according to the preceding type of probe test. Here we see again a robust effect of having shifted the reader's attention to another location, that is, the source room ($F(1,39) = 15.67, p < .001$), and to an object of varying relevance to the protagonist's goal ($F(1,39) = 41.47, p < .001$).

The effect of probe-object location here was probably due to readers being required to shift their focus of attention *back* to the current location room from which the action of the main narrative then proceeded. The effect of the goal-relevance of the preceding probe object was probably caused by a similar reshifting from the irrelevant probe object to the proper probe location implied by such phrases as "After he had finished *that task.* ..." To illustrate, suppose the goal implied use of the *Xerox machine* in the experiment (location) room, but the probe asked whether the *camcorder* is in the source room (an "irrelevant source room" probe). To answer that, participants must access the location of the *camcorder* and shift focus to the queried source room. Then the following sentence's phrase "finished *that task*" required participants to reinstate the goal (to copy notes on the

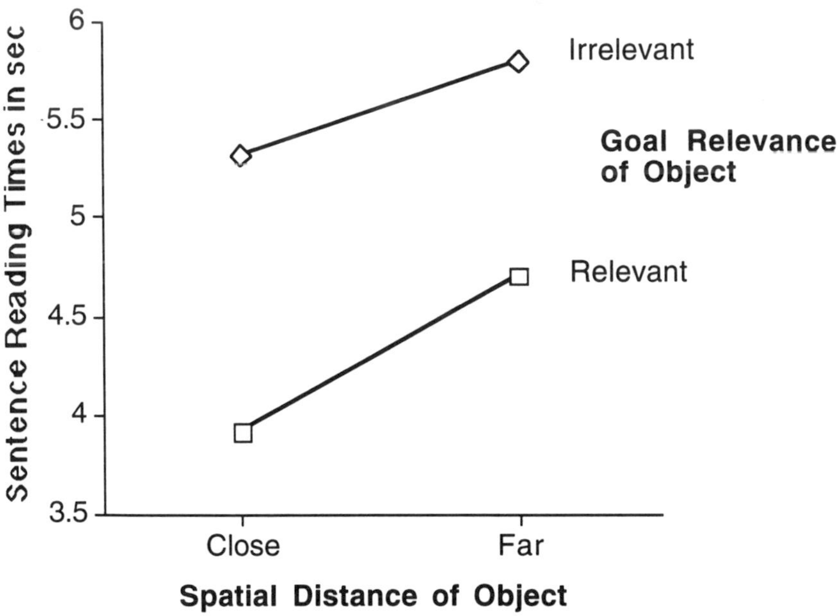

FIG. 8.8. Time to read the sentence that follows the four different object-in-room questions. From an experiment by Rinck, Bower, and Graves.

Xerox machine) and then shift attention back to the location room where the protagonist has just now finished that Xeroxing task. We may thus explain the reading-time results for the following sentences as due to implied distance-effects in shifting focus of attention within the mental model.

Do Goals Underlie Spatial Priming?

We may ask whether the spatial priming of objects in the situation model is in reality just a kind of "relevance" effect. Readers may direct their attention or internal focus to places where significant events are likely to occur. Thus, when a character moves to some location, the objects there may be activated because we would normally expect them to be relevant to the character's goal. This strategy is shaped and supported by the arrangement of human living spaces in which different parts of a building, say a residence, are set aside to satisfy recurrent goals, as revealed in our labeling of such rooms as the kitchen, bedroom, laundry, toilet, and entertainment center.

According to this speculation, then, spatial priming in narratives would be a derivative of readers' subconscious expectations that characters go to a place in order to carry out some goal-plan with the people or objects there. By this idea, we have returned to our earlier starting place, one that was foreshadowed in Tom Trabasso's earlier research on causal analyses of narratives. This is the hypothesis that relevance in narrative comprehension is largely determined by the goals and plans of the central characters. Readers use the characters' goals to draw causal connections among events and to focus attention on actors, places, and objects in the situation model that are likely to be relevant to achieving or thwarting those goals. This is as it should be: Readers are simply carrying into their story understanding the same processes and biases that they use in understanding the behaviors and plans of the social actors in their everyday lives.

ACKNOWLEDGMENTS

Preparation of this chapter was supported by NIMH Grant 1R37-MH-47575 to Gordon Bower, and by Grant Ri 600/3-3 from the German Research Foundation (DFG) to Mike Rinck. We would like to thank Erin Graves for her help in preparing and conducting the experiment on goal activation of situation model objects.

REFERENCES

Anderson, J. R. (1976). *Language, memory, and thought.* Hillsdale, NJ: Lawrence Erlbaum Associates.

Baddeley, A. D. (1990). *Human memory: Theory and practice.* Boston: Allyn & Bacon.

Bartlett, F. C. (1932). *Remembering: A study in experimental and social psychology.* Cambridge, England: Cambridge University Press.

Bower, G. H. (1982). Plans and goals in understanding episodes. In A. Flammer & W. Kintsch (Eds.), *Discourse processing* (pp. 2–15). Amsterdam: North-Holland.

Bower, G. H., & Morrow, D. G. (1990). Mental models in narrative comprehension. *Science, 247,* 44–48.

Britton, B. K., & Black, J. B. (Eds.). (1985). *Understanding expository test.* Hillsdale, NJ: Lawrence Erlbaum Associates.

Brown, A. L., & Smiley, S. S. (1977). The importance of structural units in prose passages: A problem of metacognitive development. *Child Development, 48,* 1–8.

Chafe, W. L. (1972). Discourse structure and human knowledge. In J. B. Carroll & R. O. Freedle (Eds.), *Language comprehension and the acquisition of knowledge* (pp. 41–70). Washington, DC: Winston & Sons.

Chafe, W. L. (1994). *Discourse, consciousness, and time.* Chicago: University of Chicago Press.

Foss, C. L., & Bower, G. H. (1986). Understanding actions in relation to goals. In N. E. Sharkey (Ed.), *Advances in cognitive science* (Vol. 1, pp. 95–124). Chichester, England: Ellis Harwood Ltd.

Graesser, A. C., Singer, M., & Trabasso, T. (1994). Constructing inferences during narrative text comprehension. *Psychological Review, 101,* 371–395.

Johnson-Laird, P. N. (1983). *Mental models.* Cambridge, England: Cambridge University Press.

Meyer, B. J. F. (1975). *The organization of prose and its effect on recall.* Amsterdam: North-Holland.

Morrow, D. G., Bower, G. H., & Greenspan, S. (1989). Updating situation models during narrative comprehension. *Journal of Memory and Language, 27,* 292–312

Morrow, D. G., Greenspan, S., & Bower, G. H. (1987). Accessibility and situation models in narrative comprehension. *Journal of Memory and Language, 26,* 165–187.

Rinck, M., & Bower, G. H. (1995). Anaphora resolution and the focus of attention in situation models. *Journal of Memory and Language, 34,* 110–131.

Sanford, A., & Garrod, S. (1981). *Understanding written language: Explorations of comprehension beyond the sentence.* Chichester, England: Wiley.

Schank, R. C., & Abelson, R. P. (1977). *Scripts, plans, goals, understanding.* Hillsdale, NJ: Lawrence Erlbaum Associates.

Sharkey, N. E., & Bower, G. H. (1987). A model of memory organization for interacting goals. In P. Morris (Ed.), *Modelling cognition* (pp. 231–248). Chichester, England: Wiley.

Sternberg, S. (1966). High-speed scanning in human memory. *Science, 153,* 652–654.

Trabasso, T., & Nickels, M. (1992). The development of goal plans of action in the narration of a picture story. *Discourse Processes, 15,* 249–275.

Trabasso, T., & Sperry, L. L. (1985). Causal relatedness and importance of story events. *Journal of Memory and Language, 24,* 595–611.

Trabasso, T., & van den Broek, P. (1985). Causal thinking and the representation of narrative events. *Journal of Memory and Language, 24,* 612–630.

Trabasso, T., van den Broek, P., & Suh, S. (1989). Logical necessity and transitivity of causal relations in stories. *Discourse Processes, 12,* 1–25.

van Dijk, T. A., & Kintsch, W. (1983). *Strategies of discourse comprehension.* New York: Academic Press.

Wilensky, R. (1978). *Understanding goal-based stories.* (Unpublished doctoral dissertation and Research Rep. No. 140). New Haven, CT: Yale University, Computer Science Dept.

Zwaan, R. A., & Radvansky, G. A. (1998). Situation models in language comprehension and memory. *Psychological Bulletin, 123,* 162–185.

9

Children's Understanding of Complex Stories: Issues of Representation and Assessment

Susan R. Goldman, Keisha O'Banion Varma, Diana Sharp, and the Cognition and Technology Group at Vanderbilt
Vanderbilt University

One of the enduring problems in American education is fostering the development of children's literacy skills. Many children of otherwise normal intelligence and academic ability come to the formal schooling environment with weak and underdeveloped skills in comprehending discourse (National Assessment of Educational Progress, 1995). This often places these children at risk for subsequent school failure (Juel, 1992; Slavin, 1994). Thus, a significant challenge for educators is how to design learning environments that support the development of literacy skills for all students, especially those whose initial skills place them at risk. Responding to such a challenge requires detailed knowledge of the development of children's capabilities for representing simple and complex forms of discourse, coupled with ways to adequately assess those capabilities. The work we report in this chapter attempts to address some of these key concerns by focusing on the representation and assessment of children's understanding of complex stories.

Our work is set in the larger context of the literatures on cognitive development and discourse processing, both of which have been significantly impacted by the work of Tom Trabasso. This includes his theoretical work on the nature of causal coherence in stories and the development of a model for representing causal structure. His application of that model in studies of children's comprehension and production of stories has significantly advanced our understanding of the development of these skills. In addition, his model has had a wider impact through its influence on numerous other researchers, including ourselves. Trabasso is also among those discourse psychologists who have emphasized the selection and use of multiple measures to assist in assessing understanding (e.g., Magliano & Graesser, 1991). Specifically, he has been concerned with the use of measures that place different demands on the comprehender. This approach to assessment often provides very different perspectives on the development of children's comprehension and representation skills, including their capacity to demonstrate their competence. The roots of this approach can be found in Trabasso's earlier work assessing the influence of task demands on young children's competence in logical reasoning. In the next sections we provide a brief discussion of some of the aforementioned seminal work by Trabasso on causal structure analysis and processing resource demands as a proper context for describing the logic of our own work on story comprehension.

Causal Structure Analysis

One of Tom Trabasso's major contributions to the field of discourse psychology is the recursive network model, a model for representing narratives as causal networks. In work conducted with several of his students and colleagues, Trabasso developed the recursive network transition model as a systematic method for analyzing the causal relations among states and events in narratives (Trabasso & van den Broek, 1985). The model relies on criteria of causal necessity and sufficiency for identifying and differentiating among causal, enabling, and sequential links among states and events in narratives. The model provides a representational system for expressing the relations among key structural elements of narratives, namely goals, attempts, and outcomes, within episodes as well as across episodes in multiepisode stories (Trabasso & van den Broek, 1985). Trabasso's recursive network model built upon but moved beyond the then extant systems for narrative analysis. The extant systems were largely based on "syntactic" categories such as initiating event, internal response, and goal plan. They devoted less attention to how states and events were related to one another (see Mandler & Johnson, 1977; Rumelhart, 1975; Stein & Glenn, 1979; Thorndyke, 1977).

Trabasso and colleagues demonstrated the power of this system for representing narratives, predicting importance ratings, and predicting recall in multiepisode stories. For example, the representations of sequential versus embedded relations among episodes were easily distinguishable using the recursive network model (Goldman, 1985; Goldman & Varnhagen, 1986; Trabasso & van den Broek, 1985). Trabasso and Sperry (1985) used causal analysis to provide a theoretically based explanation for importance judgments and showed that it had impressive predictive power (van den Broek, 1988). With respect to memory for stories, Nezworski, Stein, and Trabasso (1982) demonstrated that causal connections was a more powerful variable than category of information in accounting for recall.

The recursive network model was extremely generative, prompting researchers to examine differences in children's as well as adults' understanding and construction of causal structure in stories (van den Broek & Trabasso, 1986). One interesting finding was that for relatively simple and short stories (e.g., three episodes of between 8 and 15 sentences per episode), children seemed to be more prone than adults to drawing causal connections among events even when these connections were not supported by the text (Goldman & Varnhagen, 1986). This tendency was stronger in younger (8-year-olds) as compared to older (10-year-olds) students, and students who were listening to as compared to reading the text (Goldman, 1985).

The model was also quite useful in examining story production. For example, Trabasso and Rodkin (1994) found that when generating narratives to the picture story "Frog, Where Are You?," children as young as 5 years old began to organize narrations into goal-action-outcome episodes and relate multiple episodes to the same goal plan of the protagonist. Younger children of 3 and 4 years of age tended not to create global coherence, although they did try to achieve local coherence by connecting adjacent clauses.

Thus it appears that relatively early in children's lives, the idea of intentional action is a powerful mechanism for understanding and producing narratives. However, children may not be able to demonstrate their command of the causal relations inherent in intentional action because of task-related processing demands. High processing demands can hamper children's understanding of the causal connections in a story or their inclusion of causal connections in stories they generate. We elaborate on this in the next section.

Demands on Processing Resources

Earlier in his career, Trabasso investigated how task demands influence the kind of logical reasoning children are able to demonstrate (Trabasso et al., 1978). The traditional class inclusion task involves an array of objects that belong to the same class, for example, seven flowers, but that can be subdivided into subclasses, for example, four roses and three tulips. After children identify the class and subclasses, a comparative question is asked, for example, "Are there more roses or more flowers?" (Inhelder & Piaget, 1964). Trabasso et al. (1978) pointed out that the traditional class-inclusion task confronted the child with multiple perceptual and linguistic demands simultaneously. They conducted a detailed analysis of the processes required for successful performance. They found that class-inclusion performance improved when presentation formats and questioning techniques increased the likelihood that children accurately encoded the class and the subclasses. For example, children were asked to answer questions about *all* pairwise comparisons of the class and subclasses, including comparisons of the subclasses (e.g., roses to tulips, flowers to tulips, flowers to roses). One interpretation of these findings is that the formats and questioning techniques reduced processing demands by guiding children through a systematic process of inspecting and encoding the relevant features of the object arrays. Children did not have to generate their own strategy for systematic inspection and comparison. This enabled them to use processing resources to notice and partition the classes and subclasses in a way that supported accurate performance on the traditional form of the class-inclusion question.

In our own research on story understanding (Goldman, 1985; Goldman & Varnhagen, 1986), we found a discrepancy between the understanding children demonstrated in question answering as compared to retelling tasks. Children demonstrated understanding of relations between events within and across episodes when we specifically questioned these relations, even though the children did not spontaneously produce these links in their retellings. In these studies, we manipulated the processing demands of the assessment tasks taking advantage of the fact that question answering places fewer processing demands on children than retelling.

Processing demands for stories can also be manipulated on the encoding side by supporting the *construction* of a story representation. For example, Sharp et al. (1995) showed that memory performance improved when the demands of developing a representation were reduced through video-based presentation. Specifically, for some stories children saw the first part on a video; for others they heard the first part. For all stories, children only heard the last part. Recall of the last part of the story was better for stories in which the first part had been seen on video. In addition, the beginnings of the stories presented in the video condition were recalled better than the beginnings of those presented orally. We interpreted these data as indicating that the video presentation facilitated the construction of a more elaborated representation of the story to which the ending information could be "attached," probably by reducing the processing resources necessary to create the representation.

In the work we report in this chapter, we are concerned with children's understanding of complex narratives. We focus specifically on two aspects of that work: (a) methods of assessing understanding that differ in the demands they make on processing resources, specifically story retelling versus question answering, and (b) video support for constructing a mental representation of the story. We elaborate on the significance of each of these issues in the sections that follow.

Methods of Assessing Story Understanding

Story retelling is a popular method of assessing narrative understanding but it places high resource demands on children. Children must retrieve information for the story, decide which to include, how to organize it, and then how to output it. Especially for longer stories, the resource demands of producing an organized and sequentially and causally coherent story may exceed the resources available to children, especially young five- and six-year-olds. Question answering reduces resource demands in three ways. First, the question provides a retrieval path: The wording of the question contains cues that can activate related information. Under these circumstances, children may be able to access information that is in their mental representation but for which they themselves cannot generate appropriate retrieval cues. Second, answering a question reduces the amount of information that needs to be held in working memory and organized for sequential linguistic output at any one time. Finally, from a pragmatic standpoint, the question makes it clear to the child what the listener is interested in hearing about. In contrast, for a story-retelling task, the child must allocate processing resources to making decisions about listeners' needs and what to include in the story (Meyerson, Goldman, Coté, Mayfield-Stewart, & Bloome, 1998). Although our data do not allow us to distinguish between these sources of facilitation, the data reported in this chapter clearly show that question answering allows children to reveal much more of what they understand from a complex story than does story retelling.

Video as a Support for Story Understanding

A second way in which we have been attempting to facilitate children's understanding of complex stories is through video presentation. As we have argued elsewhere, video may have an important role to play in improving reading comprehension especially among children who are at-risk with respect to literacy skills (Bransford et al., 1996; Cognition & Technology Group at Vanderbilt [CTGV], 1998). Video can be a powerful mechanism for engaging children in an active comprehension process. We have demonstrated that it can be a means of bootstrapping print-based literacy skills when used as a starting point for indepth comprehension and writing activities (CTGV, 1998). This is quite different from video as "entertainment" where the viewer is little more than a passive "couch potato."

Certainly, video tools are not necessary for building good comprehension skills. However, in today's schools teachers are faced with groups of children whose preschool literacy experiences are widely disparate. Some children come from highly literate homes where book reading is a daily event; others may have had little or no experience with books and stories prior to entering the classroom (Heath, 1982; Snow & Niñio, 1986; Teale, 1986). In the face of what appear to be weak literacy skills, teachers may restrict the kinds of stories they use in class to very simple texts that do not permit rich discussions or highly meaningful interactions. This may prove counterproductive for several reasons. It may further delay students' comprehension skill development because they will not have opportunities to deal with understanding complex and multiply caused states and events. They may also lose interest in the stories they are to read and functionally "drop out" of lessons based on simple stories. Because video is a familiar medium to most children, it may provide an important and motivativing bridge to oral and print-based literacy by providing more meaningful, motivating, and complex comprehension opportunities.

Work on video support continues a line of research that has looked at how pictures affect comprehension. One major finding is that pictures facilitate comprehension if they are consistent with the verbal information in a text (Guttmann, Levin, & Pressley, 1977; Levie & Lentz, 1982; Levin, 1981). If the pictures illustrate information central to the text, new content important to the overall message, or depict structural relationships mentioned in the text then recall of text increases, particularly for poor readers (Guttmann et al., 1977; Levin, 1981). Pictures are thought to enhance comprehension by assisting in the formation of a mental representation of the information conveyed through language. Glenberg and Langston (1992) concluded from their work on the cognitive processes underlying the facilitative effects of pictures that pictures provide an advantage because they depict relationships between objects in texts, especially spatial relationships. These are relationships that are less easily comprehended or less easily kept track of when described in language alone. More accurate and complete representations of this information will support more accurate and appropriate inferences.

Video is capable of performing functions similar to those of pictures. However, because video is a richer medium than still pictures, it can be seen as a maximal form of visual mental model support. For example, video can effectively portray dynamic relationships in ways that still pictures cannot. In several studies of televised stories, researchers have found that children remember central story actions better for televised stories than for orally presented stories, even when the latter are accompanied by static pictures (Beagles-Roos & Gat, 1983; Gibbons, Anderson, Smith, Field, & Fischer, 1986; Hayes, Kelly, & Mandel, 1986; Meringoff, 1980; Pezdek, Lehrer, & Simon, 1984; Sharp et al., 1995). In addition, memory for details and vague characters is enhanced by television (Beagles-Roos & Gat, 1983).

These findings suggest that video supports the comprehension processes necessary to form a mental model. Mental models, in turn, support high-level comprehension processes, such as drawing inferences (Kintsch & van Dijk, 1978; Trabasso & Nicholas, 1980; van den Broek, 1989). Although several media comparison studies have been conducted (Beagles-Roos & Gat, 1983; Clark, 1983; Neuman, 1992), they have not approached the issue from the standpoint of causal coherence.

Causal coherence is important because comprehension involves the construction of a representation that reflects both the content of the text and the situation conveyed by the text (e.g., Coté, Goldman, & Saul, 1998; Goldman & van Oostendorp, 1999; Kintsch, 1998; van Dijk & Kintsch, 1983).[1] Building mental representations that capture the situation model as well as the explicit text typically involves making various inferences about the causal relations that hold among story states, events, and episodes.

---

[1]The situation model is sometimes referred to as a mental model (Johnson-Laird, 1983; McNamara, Miller, & Bransford, 1991) or the *interpretation* of a text as compared to the meaning (Perfetti, 1989). We use the term *situation model* in this chapter.

Trabasso's recursive network model (Trabasso & van den Broek, 1985) provides a means for representing the inferred as well as explicitly stated causal structure of a story, regardless of whether the story is presented on video, orally, or in print. It thus provides a principled means of indicating what information is central to the causal structure of the situation model and how that information is logically related to other information. Previous work on video and pictures has been difficult to interpret because of the lack of a systematic way to analyze the video and pictures with respect to the structure of the text.

In the work reported here, we were specifically interested in whether video as compared to verbal presentation of a story enhanced children's understanding of the literal and inferential causal connections among story events, and across different episodes in a complex story. We examined this issue with students traditionally considered to be at-risk for literacy development as well as students not considered to be at-risk.

CHILDREN'S UNDERSTANDING OF A COMPLEX NARRATIVE: ONE HUNDRED MILLION BILLION EYES

We developed our own video in order to investigate the role of this medium in supporting understanding of complex narratives. We based the video on themes that would have appeal to children from diverse economic backgrounds, yet would capture authentic concerns of young, elementary school-age children, especially upon entry to school. We also wanted the video story to introduce some strategic comprehension skills such as "imagining the information in print in your mind" and practicing for presentations. The video story, called "One Hundred Million Billion Eyes" (Vanderbilt's Learning Technology Center, 1991), runs for approximately 10 minutes and contains approximately 2,000 words in the soundtrack. A parse of the soundtrack produced 161 statements, corresponding to complex propositions (Kintsch, 1998), in 20 episodes. The Appendix provides a description of each episode. The text of the story can be accessed through the World Wide Web (WWW) at http://peabody.vanderbilt.edu/ctrs/ltc/goldmans/miss_emma/.

A verbal version of the video was created by having the narrator of the video make an audiotape of an augmented script. The augmentations were statements inserted into an audio version to communicate information that was shown in the video but not verbally described in the video narration.

The verbal augmentations added an additional 46 statements, italicized in the text that can be found on the WWW. The additions were based on empirical data collected in a separate study (see Sharp, Goldman, & CTGV, 1993) in which adults watched the video without sound. After viewing each scene, the adults were instructed to write a complete description of what was happening. Information frequently mentioned by the adults was included verbally in the story created for the verbal condition. These additions did not affect the causal structure of the story and preserved the same literal and inferential causal links among events and episodes.

We applied the recursive network model to the story to develop a causal structure representation.[2] The overall story structure is that of an inner story embedded in an outer story. Because of the length of the story, the nodes in the causal structure shown in Fig. 9.1 are episodes rather than individual states and events within the episodes. The numbers in parentheses within each node indicate statements from the story. As Fig. 9.1 shows, the story is represented as a setting plus 20 episodes, 14 of which comprise the outer story and 6 of which comprise the inner story. The outer story concerns Angel, the main character who is celebrating her sixth birthday. One of the neighborhood teenagers, Tony, is a videographer and a friend of Angel's (Episode 1). He wants to make a video of Angel for her birthday, as he does for all of the neighborhood children. Tony wants Angel to talk into the microphone of his video camera. Angel is afraid to do this and runs away from him (Episode 2) to Miss Emma's house. Angel helps Miss Emma water her plants and tells Miss Emma about Tony and the camera (Episodes 3 and 4). Miss Emma gives her a present for her birthday just before they leave to go and meet Derrick (Episode 5). They meet Derrick at the playground. He has a birthday note for Angel that Mr. Scott, a friend of Miss Emma's, left with Derrick (Episode 6). The three of them read Angel's birthday note (Episode 6). Then they read the note that Mr. Scott has left for Miss Emma (Episode 7). It talks about leaving a chair in the shade but Angel notices that the chair is no longer in the shade (Episode 8). Episodes 6, 7, and 8 are not actually on the causal chain of the story. They were included as links to anticipated science-related instruction on light and shadow. After they finish reading the notes, the three sit under a tree, and open Angel's present, which is a book that Miss Emma has made especially for Angel (Episode 9). The book has no pictures so Miss Emma tells Angel and Derrick that they have to use their imaginations to see the pictures in their minds. Angel and Derrick talk about what they imagine (Episode 10).

[2]We thank Paul van den Broek for assisting with this process.

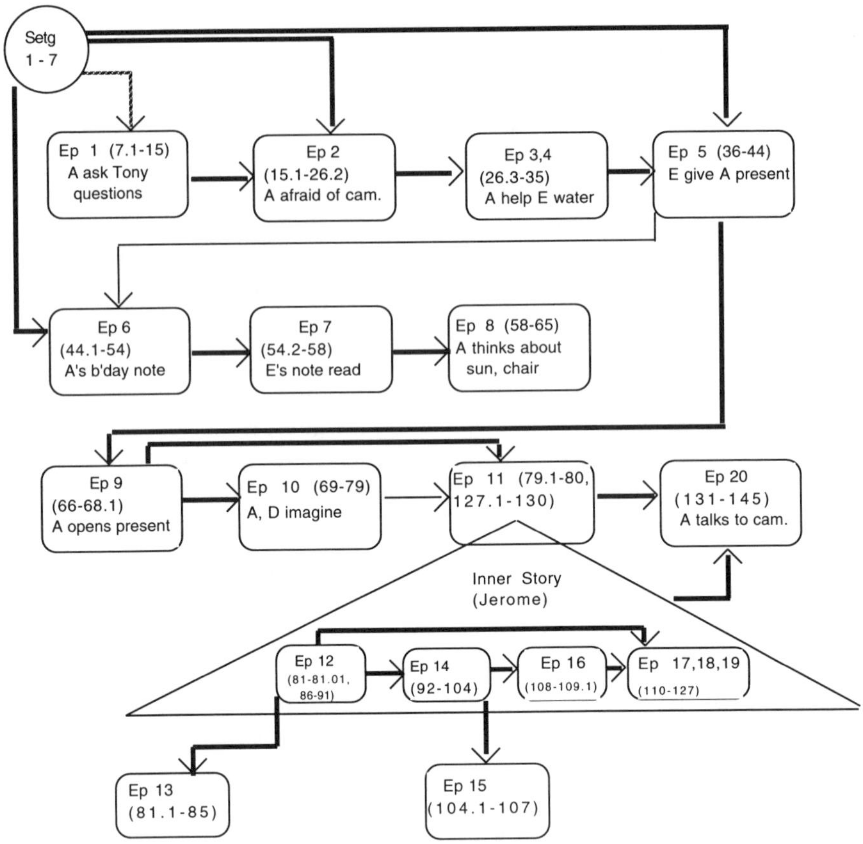

FIG. 9.1 Causal Network Diagram of the One Hundred Billion Eyes.
Causal links are indicated by ▬▬ ; enable links by ■ ■ ■ ■; and then links by ——.

Miss Emma then begins to read the book to Angel and Derrick (Episode 11), thereby embedding the inner story in the outer story.

The inner story is about a little boy, Jerome, who is afraid to speak in front of his class at show and tell (Episode 12). The inner story is interrupted by an episode in which Angel and Derrick imagine what Jerome looks like (Episode 13). Miss Emma resumes reading and Jerome talks more about his fears (Episode 12). Jerome's dad comes into his room and Jerome tells him about his fears. His dad tells him that the way to overcome a fear is to do what you are afraid of. He suggests that Jerome practice so he is ready for show and tell (Episode 14). Episode 15 pops out to the outer story when Miss Emma asks Derrick a question (Episode 15). Both episodes that leave the inner story are caused by something in the inner story but do not advance the story line. Hence they are off the causal chain. Jerome practices until it is time for school (Episode 16). Jerome makes his show-and-tell presentation at school, and finds that he can do it and that his fear has gone away; he looks forward to doing it again (Episodes 17, 18, 19). At that point, Angel and Derrick react to the story, concluding the embedding episode. Then Angel sees Tony with his video camera, and decides to talk into the microphone (Episode 20). In both the video and verbal versions, the causal connections between the inner and outer story were not explicitly stated.

We used the video and verbal versions of the story to investigate how young children understood and represented the story. Based on our earlier work with simpler narratives, we expected that video would enable youngsters to more easily build a coherent model of the story world (Sharp et al., 1992). We expected this process to be more difficult for those youngsters who had only pictures of the main characters (i. e., Angel, Miss Emma, Derrick, and Jerome) and heard the story. We were also interested in whether facilitative effects of video as compared to verbal presentation were more important for children with less well developed traditional language skills. We assessed children's understanding through prompted retellings of the story and through questions that assessed literal and inferential understanding of the story. We were especially interested in whether students understood the causal connection between the inner and outer stories. We expected that this would be more likely in the video presentation conditions, because the video support might reduce the amount of resources children needed to process the information in the story and they might be able to devote resources to making inferences about the relationships among events in the story.

Participants who were identified as being at-risk for school failure were from five kindergarten classrooms in two inner-city schools. Each scored below age level on the Peabody Picture Vocabulary Test–Revised (PPVT–R) (mean percentile = 10.25). Those who were identified as non-at-risk were from three classrooms in an elementary school located in the suburbs of the same city and scored at or above age level on the PPVT–R (mean percentile = 74.5). A total of 32 at-risk and 32 non-at-risk children participated in the study, with half of each group in the video and half in the verbal condition. The mean age for each of the four cells was 6 years 1 month. According to *t* tests, the PPVT–R scores among the at-risk children in the verbal and video conditions were not significantly different; the same was true for the non-at-risk youngsters.

Prior to hearing or seeing the story, we familiarized each child with the procedure. The child was introduced to a puppet called the *story witch.* As the puppet talked to the child, she told them that the experimenter was going to show them a movie while she took a nap. Children listened to a set of three stories ranging from seven to eight sentences. The stories used in this session were short and simple so that the children could get accustomed to the testing format. Stories used in the introductory session were accompanied by static and dynamic illustrations on videodisc. Each story had the following structure: setting, initiating event(s), goal statement, attempt, consequence, and reaction. After each practice story, the child would retell the story to the puppet and then answer some questions about the story. The questions asked about a range of information, some of which had been presented in the story (literal questions) and some of which had not been (inference questions): motive (e.g., Why did Jasper need to fix his boat?), attempt (e.g., How did he fix his boat?), emotion (e.g., How did he feel at the end of the story?), and emotion explanation (e.g., Why did he feel that way?). If children failed to answer a question, the experimenter provided a series of prompts designed to elicit the correct information and allow the child to feel successful.

In the second session, children watched or listened to *One Hundred Million Billion Eyes* and were then tested in a retelling format in which the puppet asked them to tell her the story that they just saw or heard. They were given an initial prompt, "Tell her it was a story about a girl named Angel," to begin retelling the story. Initial prompts of "And then what?" were supplied when needed to enhance recall. Transcripts were scored for the presence of the gist of the story information.

Story Retelling

Kindergarten children were able to retell only a small percentage of the story events and there was large within-cell variance. Table 9.1 provides the means and standard deviations. Gist recall of the presented story information averaged 9.8 statements in the video and 7.37 in the verbal condition; at-risk students included 7.47 statements; non-at-risk 9.72. The difference between the at-risk video (9.38) and verbal conditions (5.56) was greater than for the non-at-risk video (10.25) and verbal (9.19) conditions. Although the effects are in the predicted direction, neither the main effects nor the interaction were statistically significant.

We looked at whether children were including similar information within and across conditions. The three most frequently included statements comprise an incomplete episode: Setting: It was Angel's birthday. Internal response: She was afraid of the camera. Outcome of the story: She talked

TABLE 9.1

Number of Kindergarten Students in Each Group (Maximum = 16) Who Included Three Most Frequently Included Story Elements in Their Retellings

	At-Risk		*Non-At-Risk*	
Story Elements	*Video*	*Verbal*	*Video*	*Verbal*
It was Angel's birthday.	69%	69%	94%	75%
Angel was afraid (of camera).	50%	19%	75%	56%
Angel waved/said "Hi" to Miss Emma in the camera.	75%	56%	38%	38%

to the camera, waving and saying "Hi" to Miss Emma. Table 9.1 shows the frequencies with which each of these story elements were included by each group. The fourth most frequently included information was that Angel got a book for a present, but less than half the students in any group included this information. The data in Table 9.1 suggest that video as compared to verbal presentation led to significantly greater likelihood of inclusion of these story elements. Consistent with our expectations about the importance of video, especially for at-risk students, the frequency differences between the verbal and video were larger in the at-risk group than in the non-at-risk group.

Beyond the three story elements, different children included a variety of additional story elements. Few students included anything about Jerome and the embedded story or attempted to explain how it was that Angel was able to make her birthday video at the end of the story. We suspect that this is due to the complexity of producing a hierarchically embedded story structure. Of those children who did include information from the embedded story along with information from the outer story (three children in each group), only three preserved the embedded order in the retelling, two children in the non-at-risk video condition and one in the at-risk verbal condition. The other children first retold information from the outer story and then retold the inner story. The outer story retellings were frequently incomplete, ending with Angel receiving the book as a present; then mentioning the boy who was scared and had to do shadow puppets for show and tell. The fact that the embedding was not preserved by the majority of children who did mention both stories supports an explanation based on working-memory resources and the high demands of retelling a complex story in which one story has to be "held open" while a second story is told. Based on the question-answering data that we report later, the story-retelling task seriously underestimated what the children had understood from the story.

We were interested in whether the kinds of production constraints seen with younger children, as well as differences between at-risk and non-at-risk children and video and verbal conditions, would continue to be present in older students. O'Banion (1994) replicated the design of the kindergarten study with third-grade children. As one might expect, older children were better able to deal with the production demands of the retelling task. They included a mean of 29 statements in their retellings, with no significant differences between groups or presentation conditions. In particular, the suggestions in the kindergarten data of a video advantage for the at-risk students were not evident just from the number of statements included by the third-graders.

However, some interesting differences among the groups emerged from analyses of the information included in the third-grade retellings and the causal structures of those retellings. The most frequently included story events were similar for the at-risk video group and the two non-at-risk groups, with the at-risk verbal group showing a somewhat different pattern. Specifically, story events included most frequently (above 60% of the students) by the three similar groups were the setting (It was Angel's birthday), the initiating event that she was afraid to talk to Tony's camera, and the outcome that she indeed did talk to the camera. Although there was no one event in the inner story that was included by over 60% of the students in any of these three groups, over 60% did include some event from the inner story or a single statement summarizing the inner story (e.g., "She [Miss Emma] read a story about Jerome."). In contrast, the story elements included by more than 60% of the students in the at-risk verbal group were one from the inner and one from the outer story: Jerome was scared to go to school was included as often as the fact that it was Angel's birthday. Also noteworthy is that whereas 58% of the students in this group included the fact that Jerome did make his presentation at school, only 30% included the outcome of the outer story, that is, that Angel talked to the camera. Apparently the video presentation enabled more of the at-risk students to reproduce a story with a beginning, middle, and end as compared to the more fragmentary retellings of the at-risk verbal group.

The patterns of co-occurrence of information enabled a view of the causal structure of the retold stories. These data indicate that there was a slight advantage of the video presentation for both at-risk and non-at-risk students. Observed patterns are described in Table 9.2, along with the percentages of students in each group whose retellings fit each structure. A significantly greater percentage of students in the video presentation conditions retold the story retaining the embedding of the inner story about Jerome in the

TABLE 9.2
Structure of Retellings by Third-Graders

	At-Risk		*Non-At-Risk*	
	Video	*Verbal*	*Video*	*Verbal*
Number of students in each group	(18)	(17)	(17)	(16)
Inner story embedded in outer, explicit causal or and-then connection between them[a]	61%	23%*	71%	44%
Outer then inner, no outer story outcome	5%	35%	12%	13%
Outer story only, including outcome	5%	—	—	19%
Outer story only, no outcome	—	23%	12%	13%
Inner only	—	6%	6%	—
Fragmentary	28%	12%	—	6%

[a]Retellings were classified in this category even if the inner story was summarized. There were two ways in which students summarized the inner story: (a) Miss Emma—she read the story, or (b) one or two statements that glossed the inner story, for example, A boy was scared but did it anyway. The first method was the more likely form of summarization.
*Significantly different from video presentation condition. As well, for this category, the overall video versus verbal comparison was significant.

outer story about Angel, and completed the outer story. These students also included in their retellings a connection between the inner story and Angel's outcome actions. Many of these were "and/then" connections, but 27% of the at-risk students and 50% of the non-at-risk students who made these explicit connections made clear the *causal* connection between the inner and outer stories. In contrast, the verbal presentation groups had two dominant patterns. They either retold the outer story and then the inner story and failed to complete the outer story, or they retold only one of the stories, usually the outer (Angel) story.

The video presentation appears to have supported the creation of memory representations that enabled more complex retellings at both age levels. The benefits to the at-risk students were more substantial than to the non-at-risk students, suggesting that video may be an effective means of bootstrapping important literacy skills. However, even with video support, the retellings were incomplete for the most part. This is not surprising, because the causal structure of the story is complex and not the type of story that kindergarten students are typically asked to retell. It is somewhat more surprising that retellings by the third-graders were also very incomplete. On the other hand, a story-retelling task requires story production skills as well as comprehension skills. By using a question-answering task and comparing results to the retelling task, we were able to look at literal and inferential understanding under circumstances of reduced production load.

Understanding Versus Production Constraints

We argued earlier that the retelling task underestimates the richness and coherence of children's understanding because of the production demands for organized linguistic output, especially for a complex embedded story structure. We were interested in whether more of the story had been understood than youngsters were able to indicate in a retelling task. As Trabasso's work has shown, the language-processing demands of a task place constraints on the level of reasoning children can demonstrate (Trabasso et al., 1978). In the present context, we thought that the discrepancy might be especially large for the younger students and those who had weaker language skills. We used responses to a series of questions to gather additional information about the completeness and coherence of story understanding. In studies with adults, Graesser has consistently found that question answering extracts much more information than recall and free generation (Graesser & Clark, 1985; Graesser, Robertson, & Anderson, 1981). In previous work with children, Goldman reported a discrepancy between what was evident in free recalls of stories and the richer representational structure that was revealed by children's responses to *why* questions about goal-directed sequences in three-episode stories (Goldman, 1985; Goldman & Varnhagen, 1986).

In the studies conducted with both kindergarten and third-grade students, children were asked a series of questions. Kindergarten students were asked 21 questions, 13 of which were literal and could be answered directly from information in the story and 8 of which required students to make inferences based on the story content. The set of questions for the third-graders included literal (eight) and inference (five) questions, some of which overlapped with the kindergarten questions. However, third-graders were also asked several additional questions (four) more reflective of age-appropriate comprehension skills. In the present context, we discuss results from a subset of the literal and inference questions given to each group.

The retelling data indicated that only three kindergarten students in each of the treatment groups included information about the inner story. However, responses to questions about the inner story indicated that kindergarten students remembered a great deal more than the retellings indicated. Table 9.3 shows each of the five questions about the inner story and the frequency of children responding accurately to each one.

TABLE 9.3

Frequency of Correct Responses to Literal Questions About the Embedded Jerome Story for Kindergarten Students (n = 16 per group)

	At-Risk		*Non-At-Risk*	
	Video	*Verbal*	*Video*	*Verbal*
What did Jerome have in his hands?	100%	44%*	100%	100%
How did Jerome feel before he spoke to his dad?	50%	75%	88%	88%
What did Jerome do after his dad left the room?	81%	50%*	88%	75%
What did Jerome do to make the shadow bigger?	56%	25%*	88%	63%
Where did Jerome go after the class clapped for him?	69%	31%*	94%	63%

*Significantly different from video presentation condition.

These data clearly indicate that the non-at-risk students, for both video and verbal presentation conditions, remembered quite a bit about Jerome. Likewise, the at-risk students who had the video presentation had far better memory for the events in the inner story than their retellings indicated, especially for what he was holding and what his dad's advice prompted him to do. The at-risk students who were in the verbal presentation condition generally had the weakest performance on the inner story questions, suggesting that their retellings were closer to their understanding than was true for the other three groups.

Kindergarten students' answers to questions about the outer story also indicated that they remembered a good bit more than they included in their retellings. For example, although few students' retellings included the fact that Miss Emma's present to Angel was a book, at least 75% of the students in each of the treatment groups accurately responded to the question about this. Likewise, almost none of the kindergarten students discussed the imagination scene in their retellings, yet when specifically asked what Angel and Derrick saw when they closed their eyes, 69% or more of the students in the non-at-risk and at-risk video groups responded accurately. However, only 25% of the at-risk verbal presentation group responded correctly. On this and other literal questions, the pattern across the four treatment groups was similar to that reported for the inner story questions, with three groups (non-at-risk video and verbal, and at-risk video) responding at similar and higher levels than the at-risk verbal presentation group.

Questions about inferences also revealed that students understood the causal connections among major story events better than indicated by the retellings, especially for connections that were central to the causal goal path of the story. For example, when asked why Miss Emma gave Angel a present, 50% of the children answered correctly, with no video advantage. These data are shown in Table 9.4. Other inference questions showed some advantage to video presentation. For example, on the critical inference connection between the inner and the outer story (Why was Angel ready to talk to the camera?), more students who had the video presentation produced correct answers than those who had the verbal presentation (see Table 9.4). The video-verbal comparison was significant only among the at-risk students, although the trend was similar in the non-at-risk students. Note that none of the kindergarten students included the causal connection between the inner and outer story in their retellings.

The comparison of the kindergarten retelling and question-answering data indicate that when production demands were lowered, kindergarten students demonstrated greater memory for the presented story information as well as greater understanding of the causal structure of the story. The third-grade students generally included more of the story information in their retellings than did the kindergarten students. Consequently, we expected less of a difference between the understanding revealed by the retellings and that revealed by the question-answering task. This expectation was confirmed. Generally, there were fewer differences between the

TABLE 9.4

Frequency of Correct Responses for Kindergarten and Percentage of Correct Responses for Third Grade Students on Two Inference Questions[a]

	At-Risk		*Non-At-Risk*	
	Video	*Verbal*	*Video*	*Verbal*
Kindergarten students				
Why did Miss Emma give Angel a present?	56%	63%	88%	81%
Why was Angel ready to talk to the camera?*	44%	19%	75%	50%
Third-Grade Students				
Why was Angel ready to talk to the camera?*	61%	18%	82%	63%

[a]Percentages are based on 16 subjects per group for the kindergarten students. For the third grade the numbers of participants in each group were at-risk video $n = 18$, at-risk verbal $n = 17$, non-at-risk video $n = 17$, and non at-risk verbal $n = 16$.

*Correct responding in the at-risk verbal condition was significantly worse than in the other three groups.

retelling and question-answering task among third-graders than among kindergarten students. However, there were indications of greater understanding of the inner story when assessed through questions. Across the four groups, 17 students did not include any information about the inner story in their retellings. For the two literal questions about the inner story, 9 of the 17 students correctly answered both questions, 7 answered one or the other question correctly, and 1 answered neither correctly. Thus, there is some evidence that more of the inner story was remembered than was produced by the third-grade students.

It is interesting to note that for the literal question about an event off the causal chain (i.e., why the chair was no longer in the sun from Episode 8), correct responding was low (11% and 12% for the at-risk video and verbal groups, 31% in the non-at-risk verbal group, and 53% in the non-at-risk video group). These percentages were about the same as inclusion in the retellings for the at-risk groups, but were higher than inclusion rates among the non-at-risk students. Thus for literal information that was not central to the causal chain of the story, direct questions revealed greater memory than did the retellings, at least for some of the students. In the sample as a whole, failure to include this information may indicate its absence from the representation, very weak retrieval paths for information not on the causal chain, or an editing process operating during production. In the latter case, students may have chosen not to include it, even though they remembered it. The data do not allow us to decide among these alternatives.

Third-grade students responses to the inference questions about the relationship between the inner and outer story revealed an interesting difference from the kindergarten students. The percentage of students answering this question correctly is shown in last row of Table 9.4. Comparing the percentages for each group to the percentages of students who produced relatively complete stories during the retellings (refer to the first row in Table 9.2), we see that there is a difference of only one or two students, usually favoring the question-asking task. Thus, for the third-grade students, the production demands of the retelling task did not prevent most of them from being able to show what they understood.

For both kindergarten and third-grade students, the question-answering task revealed more information than children included in their retellings. The general pattern (video better than verbal) that was observed in the retellings was present in the question-answering task. However, the differences were not as dramatic, especially for the non-at-risk students. The at-risk verbal group was the weakest group, with performance significantly below that of the other three groups.

SUMMARY AND CONCLUSIONS

We have discussed two important issues brought to the fore by the work of Tom Trabasso on discourse comprehension. The present results illustrate the value of theoretical and methodological tools that permit the systematic and consistent analysis of stories and the representations that children construct. Causal network analysis provided us with a valuable tool for analyzing the structure of a complex story and of students' retellings. It also helped us develop questions that asked about critical relationships among events in the story. Through this process, we were able to systematically examine differences among video and verbal presentations of a story as well as similarities and differences in comprehension among children entering school with disparate literacy skills. In addition, we were able to compare performance among those just entering school with those who had attended for several years.

Causal structure analysis of the retellings provided some evidence that the video presentation compensated for the higher processing demands of the story-retelling task even among non-at-risk students. Third-graders who saw the video produced more complete and more complex causal structures in their story retellings than did students who had heard the story. This was true for at-risk and non-at-risk students. Video presentation appears to be a useful tool for reducing some of the task demands associated with processing complex stories. Students are more able to make complex inferences among events and reach a richer understanding of the story. As such, it serves as a useful way to bootstrap complex comprehension skills that support traditional literacy skills. Students can engage in complex thinking about material presented on video while they are developing fluency with print-based skills.

Second, the results demonstrate the importance of analyzing differential processing demands engendered by presentation conditions as well as by methods of assessing children's understanding. The comparison of the retelling and question-answering data showed the value of using multiple measures of understanding to draw inferences about mental representations. Especially for the younger children, the demands on working-memory resources of the story-retelling task appear to have prevented them from showing what they had understood and remembered from the story. Their answers to both literal and inferential questions indicated that they had understood a great deal more than they had retold. This was especially true for students who had viewed the story. Among third-graders, as expected, there was less of a discrepancy between the understanding of the story reflected in retellings and in the answers to the questions. Nevertheless, for at-risk students, story understanding was still quite limited in the verbal condition. Video supported more complex understanding of the relationships among story events, especially the critical inference between the inner and outer story.

In related work, we have demonstrated that beginning with complex, video-based anchor stories produces powerful effects on reading and writing (CTGV, 1998; Mayfield-Stewart & Goldman, 1998). The complexity of the event structures in such anchor stories builds on children's general comprehension of event sequences in their everyday life. The video medium appears to facilitate bridging from informal, everyday experiences to the more formal school setting in a way that predominantly verbal interactions do not. Video provides dynamic visual support that helps contextualize linguistic information in much the same way that language acquisition is contextualized in a world of visually present people, places, and objects. Indeed, a primary impetus for our work on anchored instruction in literacy, mathematics, and science is the desire to situate learning in the contexts in which the concepts and procedures are meaningful to students (CTGV, 1990, 1997). Video permits the development of comprehension of complex relationships among events without the task demands associated with processing print-based material. Resources can be used to develop skills for making inferences, particularly about causal connections among events, and developing other strategies that lead to coherent mental representations. The event-understanding and comprehension strategies that are developed in the video context can then serve as prior knowledge that may be activated and brought to bear during print-based comprehension. In this manner, video can be used to bootstrap the text-based comprehension and production skills that are vitally necessary for success in school and in the wider society.

As we noted in the introduction, one of the enduring problems of American education is fostering the development of literacy skills, including building strong connections between school and home literacy contexts. The pursuit of theories and analytical approaches, such as those developed and stimulated by Tom Trabasso, is a critical component in efforts to systematically address issues of learning, development, and assessment. Only by understanding children's capacities to understand complex discourse, and how to assess such understanding, can we take the appropriate next steps to design enhanced educational opportunities for all children.

ACKNOWLEDGMENTS

Members of the Cognition and Technology group who contributed to the work reported in this chapter are John Bransford, Maria Grisham, Ted

Hasselbring, Kathryn Ives, Jean Livernois, Cynthia Mayfield-Stewart, Elizabeth Saul, and Nancy Vye. We thank Art Graesser, James Pellegrino, and Paul van den Broek for very helpful comments on earlier drafts of this chapter. This research was funded by the National Institute of Child Health and Development, Grant P01 HD15051-12. The views expressed in this chapter should not be attributed to the funding organization.

REFERENCES

Beagles-Roos, J., & Gat, I. (1983). Specific impact of radio and television on children's story comprehension. *Journal of Educational Psychology, 75,* 128–137.

Bransford, J. D., Sharp, D. M., Vye, N. J., Goldman, S. R., Hasselbring, T. S., Goin, L., O'Banion, K., Livernois, J., Saul, E., & the Cognition and Technology Group at Vanderbilt. (1996). MOST environments for accelerating literacy development. In S. Vosniadou, E. De Corte, R. Glaser, & H. Mandl (Eds.), *International perspectives on the psychological foundations of technology-based learning environments* (pp. 223–256). Hillsdale, NJ: Lawrence Erlbaum Associates.

Clark, R. E. (1983). Reconsidering research on learning from media. *Review of Educational Research, 53,* 445–459.

Cognition and Technology Group at Vanderbilt. (1990). Anchored instruction and its relationship to situated cognition. *Educational Researcher, 19*(6), 2–10.

Cognition and Technology Group at Vanderbilt. (1997). *The Jasper project: Lessons in curriculum, instruction, assessment, and professional development.* Mahwah, NJ: Lawrence Erlbaum Associates.

Cognition and Technology Group at Vanderbilt. (1998). Designing environments to reveal, support, and expand our children's potentials. In S. A. Soraci & W. McIlvane (Eds.), *Perspectives on fundamental processes in intellectual functioning: A survey of research approaches* (Vol. 1, pp. 313–350). Stamford, CT: Ablex.

Coté, N., Goldman, S. R., & Saul, E. U. (1998). Students making sense of informational text: Relations between processing and representation. *Discourse Processes, 25,* 1–53.

Gibbons, J., Anderson, D. R., Smith, R., Field, D. E., & Fischer, C. (1986). Young children's recall and reconstruction of audio and audiovisual narratives. *Child Development, 57,* 1014–1023.

Glenberg, A. M., & Langston, W. E. (1992). Comprehension of illustrated test: Pictures help to build mental models. *Journal of Memory & Language, 31,* 129–151.

Goldman, S. R. (1985). Inferential reasoning in and about narrative texts. In A. Graesser & J. Black (Eds.), *The psychology of questions* (pp. 247–276). Hillsdale, NJ: Lawrence Erlbaum Associates.

Goldman, S. R., & Van Oostendorp, H. (1999). Conclusions, conundrums and challenges for the future. In H. Van Oostendorp & S. R. Goldman (Eds.), *The construction of mental representations during reading* (pp. 367–376). Mahwah, NJ: Lawrence Erlbaum Associates.

Goldman, S. R., & Varnhagen, C. K. (1986). Memory for embedded and sequential story structures. *Journal of Memory and Language, 25,* 401–418.

Graesser, A. C., & Clark, L. F. (1985). *Structures and procedures of implicit knowledge.* Norwood, NJ: Ablex.

Graesser, A. C., Robertson, S. P., & Anderson, P. A. (1981). Incorporating inferences in narrative representations: A study of how and why. *Cognitive Psychology, 13,* 1–26.

Guttmann, J., Levin, J. R., & Pressley, M. (1977). Pictures, partial pictures, and young children's oral prose learning. *Journal of Educational Psychology, 69,* 473–480.

Hayes, D. S., Kelly, S. B., & Mandel, M. (1986). Media differences in children's story synopses: Radio and television contrasted. *Journal of Educational Psychology, 78,* 341–346.

Heath, S. B., (1982). What no bedtime story means: Narrative skills at home and school. *Language in Society, 11,* 49–76.

Inhelder, B., & Piaget, J. (1964). *The early growth of logic in the child: Classification and seriation.* New York: Humanities Press.

Johnson-Laird, P. N. (1983). *Mental models.* Cambridge, MA: Harvard University Press.

Juel, C. (1992). Longitudinal research on learning to read and write with at-risk students. In M. J. Dreher & W. H. Slater (Eds.), *Elementary school literacy: Critical issues* (pp. 73–99). Norwood, MA: Christopher-Gordon.

Kintsch, W. (1998). *Comprehension: A paradigm for cognition.* New York, NY: Cambridge University Press.

Kintsch, W., & van Dijk, T. A. (1978). Toward a model of text comprehension and production. *Psychological Review, 85,* 363–394.

Levie, W. H., & Lentz, R. (1982). Effects of text illustrations: A review of research. *Educational Communication and Technology Journal, 30,* 195–232.

Levin, J. R. (1981). On the functions of pictures in prose. In F. J. Priozzolo & M. C. Wittrock (Eds.), *Neuropsychological and cognitive processes in reading* (pp. 203–228). New York: Academic Press.

Magliano, J. P., & Graesser, A. C. (1991). A three-pronged method for studying inference generation in literary text. *Poetics, 20,* 193–232.

Mandler, J. M., & Johnson, N. J. (1977). Remembrance of things parsed: Story structure and recall. *Cognitive Psychology, 9,* 111–151.

Mayfield-Stewart, C., & Goldman, S. (1998). *Facilitating young children's story writing with an anchored multimedia environment.* Roundtable conducted at the 1998 American Education Research Association annual meeting, San Diego, CA.

McNamara, T. P., Miller, D. L., & Bransford, J. D. (1991). Mental models and reading comprehension. In R. Barr, M. L. Kamil, P. B. Mosenthal, & P. D. Pearson (Eds.), *Handbook of reading research* (Vol. 2, pp. 490–511). New York: Longman.

Meringoff, L. K. (1980). Influence of the medium on children's story apprehension. *Journal of Educational Psychology, 72,* 240–249.

Meyerson, P. M., Goldman, S. R., Coté, N., Mayfield-Stewart, C., & Bloome, D. M. (1998, July). *"Where's Glowbird?": Children's use of multiple dimensions in story narratives.* Paper presented at the meeting of the 1998 Eighth Annual Meeting of the Society for Text and Discourse, Madison, WI.

National Assessment of Educational Progress. (1995). *NAEP 1994 reading: A first look–findings from the National Assessment of Educational Progress (Rev. Ed.).* Washington, DC: U.S. Government Printing Office.

Neuman, S. B. (1992). Is learning from media distinctive? Examining children's inferencing strategies. *American Educational Research Journal, 29,* 119–140.

Nezworski, T., Stein, N. L., & Trabasso, T. (1982). Story structure versus content in children's recall. *Journal of Verbal Learning and Verbal Behavior, 21,* 196–206.

O'Banion, K. (1994, December). *Video support for inference generation and mental model construction during story comprehension.* Unpublished master's thesis, Vanderbilt University, Nashville, TN.

Perfetti, C. A. (1989). There are generalized abilities and one of them is reading. In L. B. Resnick (Ed.), *Knowing, learning, and instruction: Essays in honor of Robert Glaser* (pp. 307–333). Hillsdale, NJ: Lawrence Erlbaum Associates.

Pezdek, K., Lehrer, A., & Simon, S. (1984). The relationship between reading and cognitive processing of television and radio. *Child Development, 55,* 2072–2082.

Rumelhart, D. E. (1975). Notes on a schema for stories. In D. Bobrow & A. Collins (Eds.), *Representation and understanding: Studies in cognitive science* (pp. 211–236). New York: Academic Press.

Sharp, D. L. M., Bransford, J. D., Vye, N., Goldman, S. R., Kinzer, C., & Soraci, S., Jr. (1992). Literacy in an age of integrated-media. In M. J. Dreher & W. H. Slater (Eds.), *Elementary school literacy: Critical issues* (pp. 183–210). Norwood, MA: Christopher-Gordon.

Sharp, D. L. M., Bransford, J. D., Goldman, S. R., Risko, V. J., Kinzer, C. K., & Vye, N. J. (1995). Dynamic visual support for story comprehension and mental model building by young, at-risk children. *Educational Technology Research and Development, 43*(4), 25–42.

Sharp, D. L. M., Goldman, S. R., & the Cognition and Technology Group at Vanderbilt. (1993, January). *Beyond action and dialogue: Video and children's story comprhension.* Paper presented at the fourth annual Winter Text Conference, Jackson Hole, WY.

Slavin, R. E. (1994). *Preventing early school failure: Research, policy, and practice.* Needham Heights, MA: Allyn & Bacon.

Snow, C. E., & Niñio, A. (1986). The contracts of literacy: What children learn from learning to read books. In W. Teale & E. Sulzby (Eds.), *Emergent literacy: Reading and writing* (pp. 116–138). Norwood, NJ: Ablex.

Stein, N. L., & Glenn, C. F. (1979). An analysis of story comprehension in elementary school children. In R. O. Freedle (Ed.), *New directions in discourse processing: Vol. 2. Advances in discourse processes* (pp. 53–120). Norwood, NJ: Ablex.

Teale, W. H. (1986). Home background and young children's literacy development. In W. H. Teale & E. Sulzby (Eds.), *Emergent literacy: Writing and reading* (pp. 173–206). Norwood, NJ: Ablex.

Thorndyke, P. W. (1977). Cognitive structures in comprehension and memory of narrative discourse. *Cognitive Psychology, 9,* 77–110.

Trabasso, T., Isen, A. M., Dolecki, P., McLanahan, A. G., Riley, C. A., & Tucker, T. (1978). How do children solve class-inclusion problems? In R. S. Siegler (Ed.), *Children's thinking: What develops?* (pp. 151–180). Hillsdale, NJ: Lawrence Erlbaum Associates.

Trabasso, T., & Nicholas, D. W. (1980). Memory and inferences in the comprehension of narratives. In F. Wilkening, J. Becker, & T. Trabasso (Eds.), *Information integration by children* (pp. 215–242). Hillsdale, NJ: Lawrence Erlbaum Associates.

Trabasso, T., & Rodkin, P. C. (1994). Knowledge of goal/plans: A conceptual basis for narrating "Frog, Where Are You?". In R. A. Berman & D. I. Slobin (Eds.), *Different ways of relating events in narrative: A cross-linguistic study* (pp. 85–106). Hillsdale, NJ: Lawrence Erlbaum Associates.

Trabasso, T., & Sperry, L. (1985). Causal relatedness and importance of story events. *Journal of Memory and Language, 24,* 595–611.

Trabasso, T., & van den Broek, P. (1985). Causal thinking and the representation of narrative events. *Journal of Memory and Language, 24,* 612–630.

van den Broek, P. (1988). The effects of causal relations and hierarchical position on the importance of story statements. *Journal of Memory and Language, 27,* 1–22.

van den Broek, P. (1989). Causal reasoning and inference making in judging the importance of story statements. *Child Development, 60,* 286–297.

van den Broek, P., & Trabasso, T. (1986). Causal networks versus goal hierarchies in summarizing text. *Discourse Processes, 9,* 1–15.

van Dijk, T. A., & Kintsch, W. (1983). *Strategies of discourse comprehension.* New York: Academic Press.

Vanderbilt's Learning Technology Center. (1991). *One hundred million billion eyes* [Video]. (Available from Young Children's Literacy Project, Box 45 Peabody, Vanderbilt University, Nashville, TN 37212)

APPENDIX

EPISODIC STRUCTURE AND TEXT OF "ONE HUNDRED MILIION BILLION EYES"

Episodic Structure of the Story as Depicted in Fig. 9.1

Setting: Angel is having her sixth birthday. 1–7.3

Episode 1: Asking questions of Tony. 8–15

Episode 2: Angel is afraid to talk to the camera. 15.1–26.2

Episodes 3, 4: Angel helps Miss Emma water plants. 26.3–35

Episode 5: Miss Emma gives Angel a present. 36–44

Episode 6: Meeting Derrick and Angel's birthday note. 44.1–54.1

Episode 7: Miss Emma reads her note. 54.2–58

Episode 8: Angel tries to figure out the sun and the chair. 58–65

Episode 9: Angel opens her present—a book. 66–68.1

Episode 10: Angel and Derrick imagine. 69–79.1

Episode 11: Miss Emma starts reading inner story. 79–80

IS: Episode 13: Jerome has flashlight, puppet; afraid to go to school. 81–81.01, 86–91

Episode 12: Angel imagines Jerome. 81.1–85.

IS: Episode 14: Dad talks with Jerome. 92–104

Inc. Episode 15: Miss Emma talks to Derrick. 104.1–107

IS: Episode 16: Jerome practices his shadow puppets. 108–109.1

IS: Episode 17: Jerome gets called on for show and tell. 110–117

IS: Episode 18: Jerome shows the puppets. 118–120

IS: Episode 19: Jerome makes the puppets bigger and smaller. 121–127

Episode 11, completion: 127.1–130

Episode 20: Angel talks to Tony's camera. 131–145

Access the full text of the story at http://peabody.vanderbilt.edu/ctrs/ltc/goldmans/miss_emma/

10

On Narrative Reading–Writing Relationships: How Young Writers Construe the Reader's Need for Inferences

Pietro Boscolo and Lerida Cisotto
University of Padova, Italy

Over the past 25 years, research on text comprehension and production has evolved along separate paths and theoretical perspectives. However, there recently have been attempts to integrate the processes of comprehension and writing. This chapter addresses the links between the reading and writing of narratives in children.

Text comprehension research has been blessed with a wealth of empirical studies and theoretical elaboration. The research was initially inspired by Kintsch's psychological models in the mid-1970s (Kintsch, 1974; Kintsch & van Dijk, 1978), and has blossomed during the 1980s and 1990s (Britton & Graesser, 1996; Gernsbacher, 1994; Graesser, Gernsbacher, & Goldman, 1997; Graesser, Millis, & Zwaan, 1997; van Dijk & Kintsch, 1983; Weaver, Mannes, & Fletcher, 1995). Text comprehension is now viewed as a complex process that coordinates multiple levels of text representation, including the surface code (i.e., wording and syntax), the rhetorical structure, the referential situation model (what the text is about), and the pragmatic communication between speaker/author and listener/reader. Investigations of narrative comprehension serve as an excellent example of the dynamics between empirical studies and theoretical elaboration. The research shifted from a focus on story structure in the 1970s (Mandler &

Johnson, 1977; Stein & Glenn, 1979; Thorndyke, 1977), to causal chains, networks, conceptual coherence, and inferences in the 1980s and 1990s (Trabasso & van den Broek, 1985; Trabasso, van den Broek, & Suh, 1989), and more recently to studies that explore the comprehension of literary and naturalistic texts (Kreuz & MacNealy, 1996; van Oostendorp & Zwaan, 1994; Zwaan, 1996).

The development of writing research started at the end of the 1970s, as a late "discovery" of the cognitive approach. Since the early 1980s, writing research has been strongly influenced by Hayes and Flower's (1980) model in which they highlighted the cognitive processes that are performed when an individual writes. They collected think-aloud protocols during the process of writing to explore how writers coordinate the expression of ideas with the constraints of language and rhetorical structure.

Researchers who investigate text comprehension and production have a number of common concerns, such as how meaning is represented, how meaning is constructed, and how the constraints of language interact with ideas. Once again, however, there has been very little systematic cross-talk between these two areas of research. To be fair, there are a few studies that have explored the relationships between reading and writing. For example, in Hayes' (1996) new formulation of Hayes and Flower's (1980) model, comprehension is assigned a relevant and important role in the writing process. In the earlier stages of writing research, McCutchen and Perfetti (1982) pointed out the gap between the two fields and attempted to fill the gap with a model of discourse coherence. Analyses of story structure have been incorporated in developmental studies of oral and written narrative production (Goldman, Cosden, & Hine, 1992; Stein, 1988; Stein & Policastro, 1984). The role of working memory has been traced in both reading and writing research (Graesser & Britton, 1996; Kellogg, 1994, 1996; McCutchen, 1996). Nevertheless, the impact of text comprehension models on writing research has been quite limited, in spite of these isolated integrations.

The lack of integration between the two research fields is exemplified in the reader-based, rather than the writer-based, perspective that is dominant in studies of discourse coherence and writing. Text comprehension research has effectively demonstrated that readers attempt to construct a coherent representation of the text they are comprehending by generating inferences and explaining why actions, events, and states are mentioned in the text (Graesser, Singer, & Trabasso, 1994; Singer, 1994; Trabasso et al., 1989; Trabasso & Suh, 1993; van den Broek, 1990). The researcher's focus is on the readers' processes, that is, on how readers construct their text representations. Analogously, in writing research, coherence is investigated not as a writer's process, but as a written product from the reader's point of view rather than the writer's point of view. Thus, the coherence of a text is usually assessed a posteriori, from the reader's perspective rather than the writer's perspective. The limitations of this approach are apparent. An analysis of how writers keep their texts coherent offers information about the text they have produced, not about the process of production.

This chapter focuses on a fundamental topic of research on reading-writing relationships: How do readers and writers interact with one another? More precisely, to what extent are young writers aware of readers' comprehension difficulties or, in text-processing terms, of the readers' need for inferences? Several studies have been conducted on this topic in recent years (Lumbelli, Paoletti, Camagni, & Frausin, 1996; Nelson & Calfee, 1998; Schriver, 1992; Tierney & Shanahan, 1991; Wong, 1991). This line of research is an excellent example of the contribution that text comprehension research can make to an integrated approach to reading-writing relationships. A model of how a reader generates inferences and explanations for a coherent text representation can also be a model of how a writer, in producing a text, anticipates a reader's understanding processes.

SOME STUDIES ON READING-WRITING RELATIONSHIPS

One reason that text comprehension research has had little impact on writing research is that the social dimension of writing has been a central focus of the sociocultural approach to literacy. According to this approach, which has noticeably contrasted the cognitive approach in the last decade, reading and writing are viewed as cultural practices. There is an emphasis on social and communicative processes, rather than cognitive processes. For example, Nystrand (1986, 1987, 1989) strongly criticized the cognitive approach to writing and instead argued for a "dialogic" relationship between reader and writer, who are both aware of each other's expectations. The basic assumption of Nystrand's (1989) socio-interactive model of writing is the reciprocity principle. According to this principle, discourse is a collaborative activity shaped by a "contract" between reader and writer. It should be noted, however, that the recent research in discourse psychology has integrated rather than contrasted the cognitive and social/communicative aspects of text comprehension (Graesser, Millis, et al., 1997; van Oostendorp & Zwaan, 1994).

Empirical studies of reader-writer interactions can be grouped into two categories. The first, influenced more by the sociocultural approach, includes studies aimed at investigating how a reader develops a sense of authorship, such as recognizing the author's intentions (Shanahan, 1998; Spivey, 1997). The second group is wider and includes studies on how young or novice writers develop a sense of audience (Kroll, 1985, 1986; Roen & Willey, 1988; Rubin, 1984). Although they follow different methodologies, these studies share some common features. Children at different school grade levels are usually asked to write, rewrite, or modify a text (e.g., a persuasive message or instructions for activating a device) in order to make it appropriate for a specific audience. The text indices generally considered are syntactic complexity, meaning changes, and stylistic features.

In more recent years, writers' awareness of readers' comprehension problems has been investigated from the standpoint of instruction. There is an attempt to improve the writers' revision abilities as a requisite for writing clearer and more understandable texts. For example, Schriver (1992) explored how to teach college students to anticipate readers' comprehension needs. If students have to develop an ability to revise for audiences other than themselves or their teachers, they must learn to consider the problems of text from the readers' point of view. Schriver designed a method called "reader-protocol teaching." This consists of giving writers feedback about how readers react to their texts, particularly when they face comprehension problems due to the poor quality of those texts. This feedback should enable writers to build a mental model of readers and their comprehension problems. Schriver's study, which was conducted on more than 100 students, demonstrated that subjects trained with the reader-protocol method became more able to diagnose readers' problems than did a control group. The reader-protocol method group was more sensitive to problems caused by gaps in written information and was more aware of problems at the global level of the text. Unfortunately, Schriver did not verify whether the participants in the reader-protocol method group also wrote clearer texts.

Lumbelli et al. (1996; Lumbelli, Paoletti, & Frausin, 1999) criticized Schriver's (1992) method in that it does not adequately consider the reader's comprehension processes. The feedback from the readers of written texts in Schriver's study consisted in pointing out the difficulties they encountered in reading (e.g., "this text needs an example," or "this section makes the idea too hard to understand"), not the processes underlying those difficulties (e.g., how a reader tries to build a representation of what he or she is reading, what elements the reader finds abstract or vague, how the reader evaluates the plausibility of inferences). Lumbelli et al.'s method was different, because they presented their participants (sixth-graders) with a reader protocol that made explicit which elaboration and inferential processes led to the solution of comprehension problems. The children were asked to read and correct texts containing unclear and missing information. During correction, the children listened to an adult's recorded comments, which included questions and inferences related to those unclear and missing parts. Thus, the protocol simulated the processes an expert reader automatically carries out in detecting a comprehension problem (e.g., filling a gap or resolving incoherence).

Both Schriver (1992) and Lumbelli et al. (1996, 1999) examined the children's ability to detect and diagnose the unclear information of a text. Lumbelli et al. also hypothesized that making comprehension problems explicit by thinking-aloud techniques had a positive influence on the children's ability to write clearer texts. Unfortunately, however, a significant correlation did not emerge between revision and text quality measures. In their hypothesis that writing may be improved by modeling the reader's dynamics of comprehension processes, Lumbelli et al. made an important step in the direction of linking writing and text-processing research. Their main concern, however, is with comprehension rather than writing. This approach essentially considers writing to be a dependent variable.

Our study also investigated children' awareness of readers' comprehension needs. Unlike other studies, however, we were not interested in ascertaining if, or how, readers' feedback can improve children's ability to write and revise. Instead, we explored how children track and monitor the audience when writing and revising a specific text type, namely narrative. Our analysis of this awareness was conducted within the framework of Graesser et al.'s (1994) constructionist theory of inference generation in narrative comprehension.

A STUDY OF NARRATIVE READING-WRITING RELATIONSHIP

The reason for our interest in narrative writing is related to the function of writing narrative in school. Children write two types of narrative in elementary and junior school: stories and narration of personal events. Story writing is generally viewed as a "free" activity by elementary school teachers. It also is an occasion for children to apply an organized story structure to the process of writing. A simplified version of the Stein and Glenn (1979) story grammar is frequently used by Italian elementary school teachers.

Whereas story writing is confined to elementary school, written narration of personal events is a frequent activity at both school levels. Several studies have analyzed the differences between children's oral production of personal event narrative and fictional stories (Allen, Kertoy, Sherblom, & Pettit, 1994; Hudson & Shapiro, 1991; Klecan-Aker, McIngvale, & Swank, 1987). In the written form, these two types of production appear to require different degrees of sensitivity to the audience or reader awareness. This may be less prevalent in story writing, where text organization is emphasized by teachers. The communicative aspects of writing are stressed more in the narration of personal experiences than in the writing of stories. Therefore, one should expect a writer to be more aware of readers' comprehension problems when narrating what really happened to him or her than when writing an invented story.

The work reported in this chapter addresses several questions. The first, and most general question addresses young writers' ability to anticipate readers' inferences. Is this ability related to the school grade level? The second question addresses the types of inferences anticipated by young writers. Specifically, is there any difference between the inferences aimed at making a text representation coherent and those inferences that elaborate the situation model? The third question addresses differences in writing and revising behavior when contrasting a true narrative versus an invented story. What differences, if any, emerge between these two genres?

We have adopted two different, but complementary, perspectives. The first focuses on the strategies a young writer spontaneously uses in order to anticipate reader's difficulties. The second perspective focuses on the degree to which a writer evaluates his or her own text as understandable. As for the first perspective, several studies (Kroll, 1985, 1986; Roen & Willey, 1988; Rubin, 1984; Tierney & Shanahan, 1991) have shown that young writers can anticipate readers' needs and inferences, although less effectively than adult ones. When writing, writers try to keep their texts coherent. Lexical choices, sentence composition, sequencing of sentences, and use of cohesion links can be regarded methods to facilitate readers' "search for meaning" (e.g., Bereiter & Scardamalia, 1987; De Weck & Schneuwly, 1994; Favart & Passerault, 1996; Pontecorvo, Orsolini, Burge, & Resnick, 1996; Scinto, 1986). Our methodological option for examining processes as well as products consisted in comparing oral "first drafts" of narrative with their subsequently produced written versions. Although there are differences between oral and written narratives, this method avoids some of the problems associated with other forms of process data (such as think-aloud protocols and error detection). We analyzed changes (spontaneous revisions) from the oral to the written version as a primary means of understanding writers' perceptions of readers' need for coherence.

Whereas the first analysis focuses on spontaneous revision (i.e. the changes writers make in writing a text composed orally), the second analysis taps the writers' evaluation of whether the written text fulfills their communicative intentions. In this case, the analysis is conducted in an interactive way. The writer's attention is drawn to specific segments of text by another person and the writer must decide if those segments are understandable or need more information. In our analysis of children's writing and revising strategies, we incorporated the categories of knowledge-based inferences in Graesser et al.'s (1994) constructionist theory of narrative comprehension. The theory assumes that readers encode three sets of inferences: (a) inferences that address the reader's comprehension goals, (b) inferences that explain why events, action, and states occur, and (c) inferences that establish coherence in the reader's situation model at local and global levels. Among the inferences listed by the authors, who make no claim of exhaustiveness, a distinction was made between (a) the explanation-based inferences that are needed to establish local and global coherence and (b) the elaborative inferences that are not needed for a coherent meaning representation. Among the former, two inferences are critical to explanation: finding a bridge between a current explicit action or event and a previous passage context (causal antecedent), and the motive for an agent's intentional action (superordinate goal). The character's emotional reactions are also important for establishing global coherence, but are not explanation based. There are many types of elaborative inferences: causal consequences (the chain of mechanical consequences of a text event, forecasted into the future) and states (i.e., a description of the physical setting, objects, characters, and psychological aspects of a situation that are not causally related to the story plot). In our research, we asked the children questions that tapped the inferences affiliated with explanation or coherence: causal antecedents, emotional reactions, superordinate goals. We also asked them questions that tap elaborative inferences: location, time, and causal consequences.

Method

Participants. The participants were 90 fourth-, sixth-, and eighth-graders, with 30 assigned to each grade level. The children were enrolled in an elementary and junior school near Padova, Italy. None of the children had any language or cognitive impairment.

Procedure. Each subject was individually asked to carry out one of two tasks, corresponding to two conditions: *invented story* versus *true story.* The instructions in the two conditions are presented below.

- *Invented story*: You should invent a short story about the following: A boy (or girl) sees an object he or she likes very much in a shop window, but he or she has not got enough money. How can he or she get that object? Construct the story orally, then write it.
- *True Story*: You must have some time seen an object you like very much in a shop window, but you do not have enough money. How do you get that object? Narrate orally, then write.

For each school grade level, 15 subjects were randomly assigned to the invented story and 15 to the true story condition.

During oral production, which was recorded, an adult was present, but did not intervene. After the oral production, each child listened to the story and was told: "This is the story you have narrated. When writing it, you can modify it (delete, add, change something) in order to make it more understandable to a reader." After the written production, each child was required to read his or her text.

Later in the session, the second author of this chapter reread the text and highlighted some points with a question. An example question is: "Here you have not written why the money was not enough. Do you think it is better to add some information to make this point clearer for a reader, or is it not necessary, and why?" The questions were the same in all the texts. Three questions tapped inferences that established explanation or coherence (Graesser et al., 1994):

1. Why was the money not enough? (causal antecedent)
2. What did the protagonist (you) feel when realizing that there was not enough money? (emotional reaction)
3. What would the protagonist (you) do with that object? (superordinate goal)

Three questions tapped elaborative inferences:

4. Where in the window or inside the shop was the object located? (state)
5. Is a specification of the setting (was it morning, afternoon, summer, etc.) necessary for text comprehension? (state)
6. When the protagonist (you) got the money, did he or she (you) spend it all to buy the object or save part of it? (causal consequence)

These questions were not asked when the information was explicit in a written text. In such instances, the question was replaced with another question of the same type (true inference or elaboration) that was justified by the text.

Results and Discussion

Two analyses were performed on the data in this study. They included (a) changes from oral to written production and (b) children's answers to adult's questions regarding the need for more information.

Children's recorded stories were transcribed and each child's oral and written productions were compared. Although the focus of our analysis was on children's sense of audience, we offer some general observations on story production, particularly regarding length and content. Regarding length, there was an expected increase from fourth- to eighth-graders' stories. Although the given plot represented a common basis for all writers, it could be developed with different numbers and types of episodes. Regarding content, in the true stories the protagonist's paths to the goal (obtaining the object desired) were rather similar. Because the object seen in a shop window was too expensive, the child either received it from his or her parents as a Christmas or birthday present, or managed to earn the necessary amount of money. For some eighth-graders, the personal narration was an occasion to describe their relationship with their parents. In many cases, the object desired became the initiating event of a conflict between generations. In the invented stories, the type of object was quite similar to the true stories: a toy or video game for younger children, a computer for older boys, clothes for younger boys and girls. The paths were more varied (from going abroad to get money for older children to a magic solution for the younger children). In the older children's invented stories, there often was a large discrepancy between the goal (obtaining a concrete and rather modest object) and the protagonist's efforts to get it (e.g., trips around the world lasting years in order to earn the money to buy a computer). We comment later on this aspect of narrative production.

Comparison of Oral and Written Productions. We compared each child's oral and written productions in an effort to identify a variety of changes. We regarded the addition and deletion of words, phrases, and sentences as quantitative changes, whereas replacements were regarded as qualitative changes. Although any changes may potentially contribute to making a text more understandable, we focused on those changes that obviously considered the reader's point of view, that is, those aimed mainly at helping a reader understand. We distinguished among these changes by adopting Graesser et al.'s (1994) taxonomy of inferences. In the transcript of each child's oral production, we underlined inferential gaps and coherence breaks because these segments trigger a reader's need for explanation (particularly, causal antecedents) to maintain coherence. We then observed whether those gaps and coherence breaks had been filled in the written version. The comparison was carried out by the two authors separately, and the results were compared and discussed. We discuss three aspects of these analyses: differences between oral and written productions, the nature of additions, and the nature of deletions.

Differences between oral and written productions increased with school grade level. Whereas fourth-graders produced oral and written stories of the same length, older children generally wrote more than they narrated orally, using oral production as a plan to be developed in the written form. This trend confirms what writing research has demonstrated (e.g., Bereiter & Scardamalia, 1987; Boscolo, 1990), namely that older children have a greater tendency to differentiate planning from writing. The difference across school grade levels was more apparent for additions. Whereas fourth-graders inserted new descriptions within specific episodes, sixth- and eighth-graders added obstacles to the goal, thus lengthening the protagonist's path to the object wanted. Eighth-graders also added personal reflections.

In fourth-graders' written productions, most additions referred to emotional reactions of the protagonist and/or other characters, and to descriptions of the desired object. Specification of the name of the toy and the protagonist's friends were also relatively frequent. Causal antecedents were occasionally produced at this school grade level, although less frequently than reactions and descriptions. Two examples are provided next:

1. Va, Grade 4 (true story):
   - Oral: ... then my mother told me she would give it [the object] to me as a present for St. Valentine's Day.
   - Written: ... then my mother, Daniela, told me she would give it to me as a present on St. Valentine's Day, my name day.
2. Gi, Grade 4 (invented story):
   - Oral: ... she ... decides to sell all her toys to get the money. And she sells everything, goes to the shop and buys the doll.
   - Written: ... she thought: "I could sell the toys of mine I do not use. They are beautiful and will earn the money which I need to buy it."

A number of other differences were found when we examined the stories in the different age groups. Sixth-graders also added emotional reactions and causal antecedents, but only few descriptions and name specifications. Eighth-graders added causal antecedents and very few emotional reactions and descriptions. However, fewer causal antecedents were made explicit in eighth-graders' texts compared with the younger children's. This result is not surprising, because older subjects' oral productions were clearer and therefore presented fewer gaps. Regarding differences between true and invented stories, causal antecedents were more frequent in fourth-graders' true stories and in sixth-graders' invented ones. No differences emerged in the eighth-graders' productions. Sixth- and eighth-graders deleted some information, particularly emotional reactions and states.

In summary, when comparing the oral and written productions, differences in school grade level emerged with respect to the addition of information that addresses coherence and elaborations. This difference was clearer in the use of elaborations and less relevant information, which was added more frequently by younger subjects. Emotional reactions, which are needed for coherence according to the constructionist theory, followed a similar trend.

Children's Revisions. Children were asked whether any information should be added to make the text segment more understandable to the reader. Whether they decided yes or no, the children gave a positive or negative answer and justified it. A content analysis was carried out on the justifications, which were grouped into the following categories:

- Positive answers:
  1. Information is added because of its importance, without any further specification or explanation (e.g., "It was important to explain more, what is written is not enough"; "It has to be explained otherwise a reader cannot understand").
  2. Information is added because of its importance in the story structure (e.g., "Time should always be specified in a narration"; "It is important to describe a character's feelings in a story").
  3. Information is added or specified in order to underline or clarify parts or relations between segments in the text (e.g., "This is the central point of the story: It has to be made explicit"; "This has to be written to make subsequent information clear").
- Negative answers:
  1. Redundant information (e.g., "The information contained in the text is enough"; "This has already been said in several parts of the text").

2. Schematic knowledge that does not need to be added because it is supplied by the reader's schematic knowledge (e.g., "Everybody knows children usually have no money"; "It is obvious children are sorry when they don't get a thing they like").
3. Information not required by the story (e.g., "This has nothing to do with the story topic"; "This does not matter").
4. Useless information that is only acceptable as detail (e.g., "This could be added, but is not important"; "The meaning of the story is not changed by adding this information").

A number of trends emerged in the data, some being quite intuitive and others being unpredictable. Positive answers were more frequent for questions about true inferences, and negative answers for questions regarding elaborations, χ^2 (1, 540) = 169.31, $p < .001$. No significant association occurred between type of inference (i.e., causal antecedent, emotional state, and superordinate goal) and positive/negative answer. Instead, a significant association emerged between type of elaborative inferences (state, time specification, and causal consequence) and positive/negative answer (χ^2 (2, 270) = 17.26, $p < .01$). Positive answers were significantly more frequent for time specification of setting, whereas negative answers were more frequent for causal consequences. Regarding positive answer justifications, general statements about the importance of information to be added were frequent in fourth-graders' answers, and tended to disappear at higher levels. The same trend was found for justifications based on text structure. Negative justifications about information irrelevance of elaborative inferences were frequent across all levels and writing tasks.

For each type of justification, subjects were assigned a percentage score that divided the number of justifications by the total number of positive (3) or negative (4) responses. A series of analyses of variance (ANOVAs) were performed on these scores as a function of grade (4, 6, vs. 8) and story type (invented vs. true). No differences emerged when comparing the total number of positive and negative justifications. There was a significant main effect of school grade level on positive justifications for information needed for specific points, with means of .22, .82, and .92 in grades 4, 6, and 8, respectively, $F(2, 83) = 77.00$, $p < .001$. There was also a significant main effect of grade level on negative justifications for redundant information, with means of .11, .20, and .28, respectively, $F(2, 83) = 5.67$, $p < .01$. However, there were no significant main effects of story type and no significant interactions between grade and story type. These results support the claim that the writer's awareness of the reader's knowledge and inferences does increase with age and this trend persists for both types of narrative.

Our study was aimed at investigating children's awareness of a reader's comprehension needs in two aspects of narrative production: the strategies a young writer adopts in trying to make his or her text understandable for a reader, and the ways he or she evaluates a reader's need for more information. The effects of the two variables on children's writing and evaluations were investigated. First, it was expected, on the basis of developmental writing research, that children's competence would increase in both solving and anticipating readers' inference. Second, we hypothesized that narrative content (i.e., personal experience vs. an invented story) could influence children's writing and evaluations. Regarding the effects of the developmental variable, our data confirm that children's ability to take a reader into account increases with school grade level. However, differences in narrative content did not have a robust effect.

The children adopted several strategies to facilitate a reader's comprehension in oral and written productions. They also adopted different attitudes for narrative writing. The fourth-graders seemed more concerned with giving information and details. The older children, although modifying their texts to a greater extent, seemed less concerned with giving information and more with enriching their texts with new episodes and personal reflections. Almost paradoxically, the eighth-graders anticipated fewer causal antecedents than did younger children during writing. This latter result should be considered in the light of two points. The first regards the higher narrative competence of older children, who produced clearer oral stories that required fewer corrections. The second consideration is that narrative writing in a school setting is rarely, if ever, framed from the standpoint of communicative. What children narrate is not read by a bona fide audience, but only by the teacher. It is therefore not surprising that older and more competent writers are more concerned with enriching their stories and less concerned with constructing a globally coherent text.

There was a different approach to narrative writing when comparing fourth-graders versus sixth- and eighth-graders. This is manifested in the children's answers and justifications to questions concerning the need to clarify specific segments of the written texts. Two results should be emphasized. The first is the attention older writers give to text features. This was apparent in the justifications related to story development, which increased with school grade. When fourth-graders identified the need for further clarification, they referred to the importance of specific information. In contrast, sixth- and eighth-graders viewed the target information as connected to other parts of the text, and emphasized the function of the target information in the structure of the story. The same trend occurred for the justifications that were based on text information. When asked to add or clarify information, older children appeared more able to present arguments that used their own texts as information sources. Thus, these two results show that older children are more careful readers of their own texts, both in connecting specific points to the global text dynamics and in referring to the text as reliable evidence.

When evaluating the need for more information about elaborative inferences, the children at all grade levels and in both production conditions expressed a generally negative, although somewhat differentiated, attitude. This "minimalist" attitude could be explained by two reasons. When writing a text type whose structure is often taught rather early in elementary school, children can differentiate clearly between what information is necessary and what is not. Detail justifications are frequent at all grade levels. But there are negative justifications for the irrelevant information as detail. Thus, when orally producing and writing a story, children construct a situation model in which essential and inessential features are clearly distinguished, although they seem to assign different degrees of relevance to the various types of elaborative inferences. For example, positive justifications were more frequent for time specifications of the setting, and negative justifications were more frequent for causal consequences.

Unfortunately, this simple account is not uniformly supported by the data that compare oral and written productions. In particular, the young writers, who rejected a reader's elaborative inferences as irrelevant, added irrelevant information to their stories. An alternative and more plausible explanation may be found in the constraints of the production task, which required the children to narrate stories about a particular plot. Such a structured task was designed to facilitate children's narrative production, as well as to compare results across the two types of narrative. Unfortunately, the "dryness" of some stories, particularly the fourth-graders', did not help individuate plausible elaborative inferences. Therefore, the sharp distinction between "true" and "useless" inferences, which emerged in the children's responses, should be verified in more free and "authentic" narrative production tasks.

This methodological consideration may also be useful in explaining the lack of major differences between true and invented stories. We expected that narrating a personal experience would encourage the writer to pay greater attention to the reader than would merely telling an invented story. However, no differences emerged from children's justifications of both positive and negative answers. We found a difference only in the specification of causal antecedents in fourth-graders' true stories, and in eighth-graders' widening of the story setting. As we have pointed out in the presentation of results, the true stories were often very similar to the invented ones, and the results may reflect this similarity. These findings indicate that it is important to put some careful thought into the design of the writing task.

Conclusions and Future Directions

At the beginning of this chapter, we acknowledged the large gulf between research on text comprehension and research on writing. We have tried to make an initial contribution toward overcoming this division by investigating the young writer's awareness of a reader's difficulties in making inferences. A theory of inference generation (Graesser et al., 1994) was incorporated as a guide in these investigations. We found this approach to be quite useful. Although the focus of the theory was on comprehension, the theory views writing and reading as reciprocal processes in the same individual, and therefore offers the possibility of an integrated approach to the study of the young writer's awareness of the reader. The theory also provides a conceptualization of inferences that serves as a valid tool for analyzing children's narrative writing behavior.

We believe that the theory could be fruitful for additional, deeper analyses of the processes that a writer goes through when planning and writing a narrative text. One direction will be to explore individual differences in writing competence. One could hypothesize that a good writer is more aware of the reader's reading goal than a poor writer. A good writer does a better job monitoring explanations and coherence. Although we did not consider the children's writing ability in our study, we did identify a developmental trend in writers' awareness of readers' needs for inferences. Another interesting research direction is suggested by the writers' differential attitudes toward elaborations versus true inferences, that is, those that are needed for establishing explanations and coherence.

ACKNOWLEDGMENTS

The preparation of this chapter was supported by Grant CNR96.05098.CT08 from the National Council of Research. The authors are particularly indebted to Susan Goldman for her suggestions and advice about the conduct and write-up of this study. They also wish to thank Art Graesser for his helpful comments and assistance in preparation of the final version of this chapter.

REFERENCES

Allen, M. S., Kertoy, M. K., Sherblom, J. C., & Pettit, J. M. (1994). Children's narrative productions: A comparison of personal event and fictional stories. *Applied Psycholinguistics, 15,* 149–176.

Bereiter, C., & Scardamalia, M. (1987). *The psychology of written communication.* Hillsdale, NJ: Lawrence Erlbaum Associates.

Boscolo, P. (1990). The construction of expository text. *First Language, 10,* 217–230.

Britton, B. K., & Graesser, A. C. (1996). *Models of understanding text.* Mahwah, NJ: Lawrence Erlbaum Associates.

De Weck, G., & Schneuwly, B. (1994). Anaphoric procedures in four text types written by children. *Discouse Processes, 17,* 465–477.

Favart, M., & Passerault, J.-M. (1996). Functionality of cohesion devices in the management of local and global coherence: Two studies in children's written production of narratives. In G. Rijlaarsdam, H. van den Bergh, & M. Couzijn (Eds.), *Theories, models and methodology in writing research* (pp. 349–365). Amsterdam: Amsterdam University Press.

Gernsbacher, M. A. (1994). *Handbook of psycholinguistics.* New York: Academic Press.

Goldman, S. R., Cosden, M. A., & Hine, M. S. (1992). Working alone and working together: Individual differences in the effects of collaboration on learning handicapped students' writing. *Learning and Individual Differences, 4,* 369–393.

Graesser, A. C., & Britton, B. K. (1996). Five metaphors for text of understanding. In B. K. Britton & A. C. Graesser (Eds.), *Models ofunderstanding text* (pp. 341–351). Mahwah, NJ: Lawrence Erlbaum Associates.

Graesser, A. C., Gernsbacher, M. A., & Goldman, S. R. (1997). Cognition. In T. A. van Dijk (Ed.), *Discourse as structure and process: Vol. 1. Discourse studies: A multidisciplinary introduction* (pp. 292–319). London: Sage.

Graesser, A. C., Millis, K. K., & Zwaan, R. A. (1997). Discourse comprehension. *Annual Review of Psychology, 48,* 163–189.

Graesser, A. C., Singer, M., & Trabasso, T. (1994). Constructing inferences during narrative text comprehension. *Psychological Review, 101,* 371–395.

Hayes, J. R. (1996). A new framework for understanding cognition and affect in writing. In C. M. Levy & S. Ransdell (Eds.), *The science of writing: Theories, methods, individual differences, and applications* (pp. 1–27). Mahwah, NJ: Lawrence Erlbaum Associates

Hayes, J. R., & Flower, L. S. (1980). Identifying the organization of writing processes. In L. W. Gregg & E. R. Steinberg (Eds.), *Cognitive processes in writing* (pp. 4–30). Hillsdale, NJ: Lawrence Erlbaum Associates.

Hudson, J. A., & Shapiro, L. R. (1991). From knowing to telling: The development of children's scripts, stories, and personal narratives. In A. McCabe & C. Peterson (Eds.), *Developing narrative structure* (pp. 89–136). Hillsdale, NJ: Lawrence Erlbaum Associates.

Kellogg, R. T. (1994). *The psychology of writing.* Oxford, England: Oxford University Press.

Kellogg, R. T. (1996). A model of working memory in writing. In In C. M. Levy & S. Ransdell (Eds.), *The science of writing: Theories, methods,individual differences, and applications* (pp. 57–71). Mahwah, NJ: Lawrence Erlbaum Associates.

Kintsch, W. (1974). *The representation of meaning in memory.* Hillsdale, NJ: Lawrence Erlbaum Associates.

Kintsch, W., & van Dijk, T. A. (1978). Towards a model of text comprehension and production. *Psychological Review, 85,* 363–394.

Klecan-Aker, J. S., McIngvale, G. K., & Swank, P. (1987). Stimulus considerations in narrative analysis of normal 3rd grade children. *Language and Speech, 30,* 13–24.

Kreuz, R. J., & MacNealy, M. S. (Eds.). (1996). *Empirical approaches to literature and aesthetics.* Norwood, NJ: Ablex.

Kroll, B. M. (1985). Rewriting a complex story for a young reader: The development of audience-adapted writing skills. *Research in the Teaching of English, 19,* 120–139.

Kroll, B. M. (1986). Explaining how to play a game: The development of informative writing skills. *Written Communication, 3,* 195–218.

Lumbelli, L., Paoletti, G., Camagni, C., & Frausin, T. (1996). Can the ability to monitor local coherence in text comprehension be transferred to writing? In G. Rijlaarsdam, H. van den Bergh, & M. Couzijn (Eds.), *Effective teaching and learning of writing* (pp. 207–223). Amsterdam: Amsterdam University Press.

Lumbelli, L., Paoletti, G., & Frausin, T. (1999). Improving the ability to detect comprehension problems: From revising to writing. *Learning and Instruction, 9,* 143–166.

Mandler, J., & Johnson, N. (1977). Remembrance of things parsed: Story structure and recall. *Cognitive Psychology, 9,* 111–151.

McCutchen, D. (1996). A capacity theory of writing: Working memory in composition. *Educational Psychology Review, 8,* 299–325.

McCutchen, D., & Perfetti, C. A. (1982). Coherence and connectedness in the development of discourse production. *Text, 2,* 113–139.

Nelson, N., & Calfee, R. C. (Eds.). (1998). *The reading-writing connection. The Ninety-Seventh Yearbook of the National Society for the Study of Education. Part 2.* Chicago: National Society for the Study of Education.

Nystrand, M. (1986). *The structure of written communication: Studies of reciprocity between writers and readers.* Orlando, FL: Academic Press.

Nystrand, M. (1987). The role of context in written communication. In R. Horowitz & S. J. Samuels (Eds.), *Comprehending oral and written language* (pp. 197–214). San Diego: Academic Press.

Nystrand, M. (1989). A social interactive model of writing. *Written Communication, 6,* 66–85.

Pontecorvo, C., Orsolini, M., Burge, B., & Resnick, L. B. (Eds.). (1996). *Children's early text construction.* Mahwah, NJ: Lawrence Erlbaum Associates.

Roen, D. H., & Willey, R. J. (1988). The effects of audience awareness on drafting and revising. *Research in the Teaching of English, 22,* 75–88.

Rubin, D. (1984). Social cognition and written communication. *Written Communication, 1,* 211–246.

Schriver, K. (1992). Teaching writers to anticipate readers' needs. *Written Communication, 9,* 179–208.

Scinto, L. F. M. (1986). *Written language and psychological development.* Orlando, FL: Academic Press.

Shanahan, T. (1998). Reader's awareness of author. In N. Nelson & R. C. Calfee (Eds.), *The reading-writing connection. The Ninety-Seventh Yearbook of the National Society for the Study of Education. Part 2* (pp. 88–111). Chicago: National Society for the Study of Education.

Singer, M. (1994). Discourse inference processes. In M. A. Gernsbacher (Ed.), *Handbook of psycholinguistics* (pp. 479–516). New York: Academic Press.

Spivey, N. N. (1997). *The constructivist metaphor.* San Diego: Academic Press.

Stein, N. L. (1988). The development of children's storytelling skill. In M. B. Franklin & S. S. Barten (Eds.), *Child language: A reader* (pp. 282–297). New York: Oxford University Press.

Stein, N. L., & Glenn, C. G. (1979). An analysis of story comprehension in elementary school children. In R. Freedle (Ed.),Advances in discourse processes: vol. 2. *New directions in discourse processes, 11* (pp. 53–120). Norwood, NJ: Ablex.
Stein, N. L., & Policastro, M. (1984). The concept of story: A comparison between children's and teacher's viewpoints. In H. Mandl, N. L. Stein, & T. Trabasso (Eds), *Learning and comprehension of text* (pp. 113–159). Hillsdale, NJ: Lawrence Erlbaum Associates.
Thorndyke, P. W. (1977). Cognitive structures in comprehension and memory of narrative discourse. *Cognitive Psychology, 9,* 71–110.
Tierney, R. J., & Shanahan, T. (1991). Research on the reading-writing relationship: Interactions, transactions, and outcomes. In R. Barr, M. L. Kamil, P. B. Mosenthal, & P. D. Pearson (Eds.), *Handbook of reading research* (Vol. 2, pp. 246–280). New York: Longman.
Trabasso, T., & Suh, S. Y. (1993). Understanding text: Achieving explanatory coherence through on-line inferences and mental operations in working memory. *Discourse Processes, 16,* 3–34.
Trabasso, T., & van den Broek, P. W. (1985). Causal thinking and the representation of narrative events. *Journal of Memory and Language, 24,* 612–630.
Trabasso, T., van den Broek, P. W., & Suh, S. Y. (1989). Logical necessity and transitivity of causal relations in stories. *Discourse Processes, 12,* 1–25.
van den Broek, P. (1990). Causal inferences in the comprehension of narrative text. In A. C. Graesser & G. H. Bower (Eds.), *The psychology of learning and motivation: Inferences and text comprehension* (Vol. 25, pp. 175–194). San Diego: Academic Press.
van Dijk, T. A., & Kintsch, W. (1983). *Strategies of discourse comprehension.* New York: Academic Press.
van Oostendorp, H., & Zwaan, R. A. (Eds.). (1994). *Naturalistic text comprehension.* Norwood, NJ: Ablex.
Weaver, C. A. III, Mannes, S., & Fletcher, C. R. (Eds.). (1995). *Discourse comprehension: Essays in honor of Walter Kintsch.* Hillsdale, NJ: Lawrence Erlbaum Associates.
Wong, B. Y. L. (1991). Connections between reading and writing processes. In A. McKeough & J. L. Lupart (Eds.), *Toward the practice of theory-based instruction* (pp. 66–91). Hillsdale, NJ: Lawrence Erlbaum Associates.
Zwaan, R. A. 1996). Toward a model of literary comprehension. In B. K. Britton & A. C. Graesser (Eds.), *Models of understanding text* (pp. 241–255). Mahwah, NJ: Lawrence Erlbaum Associates.

11

Detecting Causal Inconsistencies in Scientific Text

Murray Singer and Nathalie Gagnon
University of Manitoba

This study was designed to inspect readers' ability to infer causal connections in expository text. These inferences, termed *causal bridging inferences,* have predominantly been studied with reference to narratives (e.g., Keenan, Baillet, & Brown, 1984; Myers, Shinjo, & Duffy, 1987; Potts, Keenan, & Golding, 1988; Singer & Ferreira, 1983; Trabasso & Sperry, 1985). Evidence stemming from both online and memory measures of narrative comprehension has converged on the conclusion that causal bridging inferences routinely accompany comprehension. Our own studies of these issues have been guided by two levels of theoretical analysis. The higher level analysis is the constructionist framework of Graesser, Singer, and Trabasso (1994). Two central principles of constructionism bear on causal bridging inferences. First, readers are proposed to strive to preserve text coherence at the surface, textbase, and situation model levels of text representation. As a corollary, connections are proposed to be identified among the elements of the causal situation model derived from text. Second, according to the *explanation assumption,* readers routinely access the causes of physical outcomes and the reasons for intentional actions described in text. Causal outcomes and intentional actions both signal changes of state (Cheng & Holyoak, 1985; Graesser, 1981; Trabasso, van den Broek, & Suh, 1989), a feature that we posit to contribute to the initiation of causal search.

At a more detailed level of analysis, bridging inference processes have been investigated in the framework of the "inference validation model" (Singer & Halldorson, 1996; Singer, Halldorson, Lear, & Andrusiak, 1992).

According to the validation model, tentative bridging inferences, causal and otherwise, must be subjected to validation before they are endorsed by the understander. The validation model postulates a series of processes. Consider a text sequence with a potentially causal thrust, such as *The hikers were ready to cook, They picked up some wood around the tent site.* The reader must ultimately understand that the plan to cook motivates the need for a fuel, namely, the wood. According to the model, the reader first derives *from the text* an idea that mediates the causal link between the sentences. For this example, the mediating idea might be "The hikers wanted to cook with wood." If the comparison of the mediating idea and one's general knowledge reveals that idea to be accurate (e.g., wood is a fuel), then the inference has been validated. These processes by no means guarantee the validation of a candidate cause. Consider the sequence, *Dorothy poured the gasoline on the bonfire, The fire went out.* I presented data indicating that readers detect an inconsistency between these two sentences. This outcome was interpreted to reflect validation processing (Singer, 1993).

Whereas existing evidence favors the validation of causal bridging inferences in narrative comprehension, Noordman, Vonk, and Kempff (1992) reported that readers of challenging expository texts did not engage in such processing. There are at least two systematic differences between narratives and expositions that might account for this difference. First, expository text is much less predictable in its structure than the narrative (Miller, 1985). Second, expository text frequently presents information unfamiliar to the reader. The unfamiliarity of text content could readily prohibit the construction and validation of bridging inferences (Goldman, 1997). For example, an individual unaware of the inflammability of wood would fail to grasp the causal relations underlying *The hikers were ready to cook, They picked up some wood around the tent site.*

In a recent study, Singer, Harkness, and Stewart (1997) set out to reconcile the discrepant findings distinguishing narrative and expository bridging inference processing. The rationale was that the conscientious reader ought to be able to compute certain bridging inferences even for texts of unfamiliar content. Consider the sentence, *Some fireworks emit crimson flames because they contain calcium salt.* Even without knowledge of the properties of calcium salts, a reader should be able to derive the mediating fact that calcium salts produce crimson flames. Although no pertinent knowledge would be available to validate this fact, the reader might accept its accuracy, on the assumption that the writer was being cooperative (Grice, 1975).

Using the paradigm of Noordman et al. (1992), Singer et al. (1997) studied bridging inference processing for 12 science passages of Noordman et al. and for new passages, derived from a science encyclopedia. Four experiments replicated Noordman et al.'s noninference findings for their original materials. However, the results for the new science passages of Singer et al. conformed with the pattern that, according to Noordman et al., indicates that bridging inferences have accompanied comprehension. The different findings were attributed to the greater difficulty of the Noordman passages. That is, the participants needed more time to read critical target sentences in the Noordman materials than the Singer materials, and made many more errors on Noordman's experimental questions. Only when the participants were prompted to read more quickly did the data indicate that the crucial inferences were not drawn for the Singer materials (Singer et al., 1997, Experiment 3). However, it was not possible to link the extra difficulty of the Noordman materials to specific characteristics such as familiarity or propositional complexity (Singer et al., 1997, Table 4).

In this study, we focused on a different index of the reader's computation of causal bridging inferences: namely, the detection of inconsistencies between these inferences and explicit text ideas. In their own investigation of this problem, Noordman et al. (1992, Experiments 3 and 5) reported that only when readers were advised to monitor texts for contradictions did they draw causal bridging inferences that exposed the contradictions. The present aim was to show that, even in the absence of special instructions, readers of expository text detect causal inconsistencies.

Our approach can be discussed with reference to the sentence, *Liquid hydrocarbons are the most commonly used lubricants because they effectively remove heat.* This sentence appears as the "remove" version of sentence (1c) in Table 11.1. To grasp this sentence, the reader must determine how the cause (liquid hydrocarbons remove heat) accounts for the effect (liquid hydrocarbons are common lubricants). Following the validation model, complete processing of this sentence would result in an inferred mediating idea such as "lubricants should remove heat." The "remove" version of (1c) was constructed to be consistent with (1b), *Lubricants should be good coolants,* which presents an (accurate) explicit assertion about lubricants. Consider next the "add" version of (1c), *Liquid hydrocarbons are the most commonly used lubricants because they effectively add heat.* In contrast with the "remove" version, this version is posited to yield an inference inconsistent with (1b), such as "lubricants should add heat." The detection of the latter inconsistency should cause reading time for the "add" version of sentence (1c) to be longer than for the "remove" version, which is our main prediction. This predicted result should occur only if readers derive causal bridging inferences from (1c).

TABLE 11.1

Sample Sentence Sets From the Singer (1) and Noordman (2) Materials

(1)	a.	A lubricant is used to prevent contact of parts in motion and thereby reduce friction and wear. (leading)
	b.	Lubricants should be good coolants. (explicit)
	c.	Liquid hydrocarbons are the most commonly used lubricants because they effectively (remove/add) heat. (target)
	d.	A lubricant may also be used to prevent rusting of close fitting parts. (final)
	e.	Are lubricants good coolants? (experimental question)
	f.	Do lubricants prevent rusting of close fitting parts? (detail question)
(2)	a.	Skiers of varying abilities need different equipment. (leading)
	b.	It is easier to turn on the shorter compact skis. (explicit)
	c.	Beginning skiers are frequently advised to use (compact/longer) skis because beginners usually have difficulty changing direction. (target)
	d.	Once they are confident that they can control their movements, the novices can quickly advance in skill. (final)
	e.	Is it less difficult to change direction on compact skis? (experimental question)
	f.	Can novice skiers quickly advance in skill? (detail question)

MATERIALS AND PREDICTIONS

We used the experimental paradigm introduced by Noordman et al. (1992) to evaluate the prediction of a consistency effect in reading time. Following this plan, experimental passages either include the explicit sentence (1b), the *explicit* condition, or exclude (1b), the *implicit* condition. As discussed in the previous section, in the explicit condition, an inconsistency effect in reading time for target (1c) will result only if readers draw the inferences that link the clauses of (1c). In the implicit condition, in contrast, no reading time consistency effect is predicted. This is because, in the absence of the explicit sentence, there is no text idea with which an inference from the target sentence may conflict. If the readers had previous familiarity with these ideas, then reading time for the inconsistent target would exceed that for the consistent target in the implicit condition, because of the inaccuracy of the inconsistent version. However, norms presented by Singer et al. (1997) revealed that our participant population has little familiarity with the crucial facts. To summarize, the predicted pattern of target reading times constitutes an interaction between "relation" (explicit, implicit) and consistency.

Table 11.1 shows that the *lubricant* passage included a leading sentence and a final sentence. For each passage, there was also a question, which probed the idea expressed by the explicit sentence (and the hypothetical mediating idea of the target sentence). The participants' answers and answer times for these questions also bear on the main hypothesis. Consider first the proportion of affirmative answers, termed *agreement rate.* If readers do not derive bridging inferences from the target sentences, then only the explicit sentences would provide information pertinent to the questions. In that event, agreement rate would simply be higher in the explicit condition than in the implicit condition. The picture is much different if the target sentences yield bridging inferences. In the explicit-inconsistent condition, the reader would encounter one fact that supported the question statement and one that contradicted it. In the implicit-inconsistent condition, furthermore, the target sentence would yield only a single inference contradictory of the question. Agreement rate then would be lower in the implicit-inconsistent condition than the explicit-inconsistent condition. There is no compelling reason to predict a comparable difference in the consistent conditions. As was predicted for reading times, a Relation × Consistency interaction would emerge.

Finally, Noordman et al. (1992) presented an analysis of the interpretation of *answer times* for *consistent* sequences in the present paradigm. If, during comprehension, the reader fails to extract the critical bridging inference from the target sentence, then answer time for question (1e) should be longer in the implicit condition than the explicit condition. This is because the explicit sentence, present in the explicit condition only, actually presents the fact queried by question (1e). In contrast, if readers derive the crucial inference from the target sentence, then the presence of the explicit sentence in the passage should have little or no impact on the answer time for (1e); answer time should be about equal in the explicit and implicit conditions.

Noordman et al.'s (1992) logic for the answer times does not address our inconsistent conditions. In those conditions, the conflict between the target sentence and the question obscures the interpretation of the answer times.

In Experiment 1, the passages were presented as illustrated in Table 11.1. With this arrangement, the propositions derived from sentence (1b) would likely reside in working memory when (1c) is encountered. In Experiment 2, a modification of the method was designed to delete the propositions of (1b) from working memory before sentence (1c) was encountered: In particular, two sentences were interposed between (1b) and (1c). The latter arrangement was intended to determine whether, even in expository texts, readers can detect the relationship between distal sentences that convey related ideas.

We considered the possibility that, in Experiment 2, the predictions concerning target reading times and agreement rate might be decoupled. Because two sentences intervened between the explicit sentence and the target sentence, it seemed conceivable that the readers would derive a causal mediating fact from the *target sentence* but fail to notice its relation to the explicit sentence. In that circumstance, the agreement rates of Experiment 2 would resemble those of Experiment 1, but the target reading times of Experiment 2 would yield no effects of theoretical significance.

METHOD

Participants

The participants were male and female students of introductory psychology at the University of Manitoba. All participants were native speakers of English and received course credit for their participation. There were 70 participants in Experiment 1 and 73 participants in Experiment 2.

Materials

Twenty-four experimental passages were constructed from those of Singer et al. (1997, Experiment 2). Twelve of Singer et al.'s passages were those of Noordman et al. (1992), and 12 were derived from a pool of 35 that Singer et al. extracted from a science encyclopedia. In addition, 4 practice passages and 20 filler passages were chosen randomly from Singer et al.'s pool. Like Singer et al., we treated passage set (Singer, Noordman) as a variable in the design.

The experimental passages of Experiment 1 are illustrated by the two sets in Table 11.1, one Singer passage and one Noordman passage. The experimental passages were cycled across four conditions, representing the crossing of relation (explicit, implicit) and consistency. Passages were explicit in the presence of the (b) sentence, and were implicit in its absence. Likewise, passages were consistent when they included the word before the slash in the parentheses of the (c) sentence (e.g., *remove* in 1c), and inconsistent when they used the word after the slash (*add*).[1] The crucial facts implied by the consistent versions of the target sentences (and queried by the postpassage experimental questions) had modest mean familiarity scores of .30 and .32 for the Singer and Noordman materials, respectively, on a scale of -1 to 1, according to Singer et al. (1997).

[1]We manipulated consistency by varying the target sentence rather than the explicit sentence (e.g., *Lubricants should be good/poor coolants*) in order that the explicit sentence and experimental question always express the same central idea. The manipulation of the target sentences resulted in their having mean syllable lengths of 25.0 and 24.6 in the consistent and inconsistent conditions, respectively.

For Experiment 2, the leading and final sentences, in that order, were repositioned between the explicit and target sentences. In the explicit condition of Experiment 2, for example, passage (1) consisted of (1b)-(1a)-(1d)-(1c) from Table 11.1. This procedure had the merit of holding the content of the passages constant between the experiments.

For each experimental passage, there were two questions. Experimental questions, such as (1e), probed the idea that was posited to mediate the clauses of the target sentence *in the consistent condition.* The other question probed a detail of the final sentence. The detail question for the *lubricant* passage was (1f), *Do lubricants prevent rusting of close fitting parts?* The detail question had the correct answer "yes" for 50% of the passages.

The 20 filler passages were each followed by one question about pertinent knowledge that, in contrast with the experimental passages, had the correct answer "no." This was intended to discourage the participants from thoughtlessly answering "yes" to general-knowledge questions. The filler passages were not cycled across conditions, but they were randomly assigned in equal numbers to bear the form of each of the four experimental conditions. Like the experimental passages, the filler passages were followed by a detail question as well as a knowledge question. Half of the filler-detail questions had the correct answer, "yes." In all other respects, the filler passages were similar to the experimental passages.

Four counterbalanced test lists were constructed. In the first list, one fourth of the experimental passages of each set (Singer, Noordman) were randomly assigned to each of the four experimental conditions: explicit consistent, implicit consistent, explicit inconsistent, and implicit inconsistent. Each of these items was randomly assigned to a list position, subject to two restrictions: First, no more than two consecutive items were from the same Set × Relation × Consistency condition. Second, at least one item in each condition (of the total of three) appeared in each half of the list. Likewise, at least two of the five fillers that mimicked each experimental condition were assigned to each half of the list. The three remaining lists were constructed by cycling the experimental passages across conditions, using a Latin-square procedure. Each list was preceded by four practice passages, one for each of the Relation × Consistency conditions.

Procedure

The sessions were conducted with groups of one to four participants, who were tested at computer-controlled stations in separate rooms. Each participant was randomly assigned to view one of the four experimental lists. The materials were presented on monochrome video monitors at a distance of 22 cm from the participant, and responses were made using the "x" and "." keys of the computer keyboards.

Upon receiving a signal, the participant read the passage in a self-paced manner, pressing the keyboard space bar to signal the understanding of the current sentence and to view the next one. Each successive sentence was presented 50 ms after the space bar was pressed, and all prior sentences of the text remained on the screen. Ten seconds after the presentation of a new sentence, the next sentence was added if the participant had not yet responded. All sentences were left-adjusted on successive lines of the screen, beginning with Line 5. When the participant signaled the understanding of the last sentence of a passage, a blank screen appeared for 3 s. Then, a fixation point was presented for 500 ms, followed by a question. There was a 6.0-s answer time limit that, if exceeded, resulted in the trial being treated as though the participant had answered "no." After a 2.5-s interval, the second question was presented, with the knowledge and detail questions for a passage presented in random order. Three seconds after the second question was answered, the participant was prompted to initiate the next trial. The "x" and "." keys were randomly assigned the labels *yes* and *no,* or vice versa. Reading times and answer times were automatically recorded to centisecond and millisecond accuracy, respectively. There was a 40-s rest period halfway through the list, disregarding the practice passages.

RESULTS AND DISCUSSION

The results are organized with reference to the different dependent variables. For each variable, the outcomes of both experiments are presented. A criterion of $\alpha = .05$ was used for all of the analyses, unless otherwise noted.

Target Reading Times

The reading times of Experiment 1 are shown in Table 11.2. Analysis of variance (ANOVA) revealed a Consistency × Relation interaction, $F_1(1,66) = 15.8$, $MSE = 9,010$, $F_2(1,22) = 8.99$, $MSE = 2,062$. This reflected that the reading time difference between inconsistent and consistent targets was greater in the explicit condition than the implicit condition. Tests of simple main effects indicated that reading times were longer for inconsistent than consistent targets in the explicit condition, $F_1(1,66) = 33.0$, $MSE = 14,215$, $F_2(1,22) = 6.59$, $MSE = 3,485$, but not in the implicit condition, $F\text{s} \leq 3.16$.

TABLE 11.2

Mean Reading Time (in Centiseconds) for Target Sentences in Experiments 1 and 2

		Explicit		*Implicit*	
Experiment	*Set*	*Consistent*	*Inconsistent*	*Consistent*	*Inconsistent*
Experiment 1	Singer	499	571	500	523
	Noordman	509	586	531	547
Experiment 2	Singer	502	549	509	544
	Noordman	544	594	547	553

In Experiment 1, target reading time was longer in the inconsistent condition than the consistent condition, $F_1(1,66) = 25.5$, $MSE = 13{,}723$, $F_2(1,22) = 12.2$, $MSE = 4{,}357$. Reading time was 20 cs longer for Noordman than Singer targets, but this effect reached significance only in the participants-random analysis, $F_1(1,66) = 4.46$, $MSE = 14{,}528$.[2]

In Experiment 2, eight participants made no affirmative responses in one or more experimental conditions, and the data were missing for these cells. The target reading times, based on the performance of the remaining 65 participants, appear in Table 11.2. The Relation × Consistency interaction was significant in the participants-random analysis, $F_1(1,61) = 5.36$, $MSE = 13{,}546$; and approached significance with items-random, $F_2(1,22) = 2.67$, $MSE = 1{,}808$, $p = .12$. Furthermore, tests of simple main effects revealed that inconsistent reading times exceeded consistent reading times in the explicit condition, $F_1(1,61) = 11.7$, $MSE = 17{,}901$, $F_2(1,22) = 9.95$, $MSE = 2{,}794$; but not in the implicit condition, $Fs \leq 2.04$. No other effects reached significance in both the participants- and items-random analyses.

Following the validation model, the reading time results are explained as follows: In the explicit condition, readers encode the explicit sentence, such as *Lubricants should be good coolants* (see Table 11.1). Second, comprehending the *target* sentence yields a mediating idea, such as "effective lubricants remove heat" or "effective lubricants add heat" in the consistent condition and the inconsistent condition, respectively. Third, the mediating idea is readily compared with the corresponding explicit sentence. Detecting an inconsistency between the mediating idea and the explicit sentence (e.g., *Lubricants should be good coolants* vs. "effective lubricants add heat") is proposed to inflate processing time.

---

[2]In addition to the effects theoretical significance, the participants-random ANOVAs throughout this study revealed interactions involving the List variable, although not main effects of List. These interactions appear because factors such as sentence length and word frequency are not controlled between lists. However, these interactions do not compromise the other effects, because control of the theoretically important variables is achieved by cycling each passage across experimental conditions in the four lists. Due to space limitations, the list interactions are not presented.

In the implicit condition, in contrast, the consistency effect was not significant. This outcome is attributed to the fact that, in the absence of the explicit sentence, the "inconsistent" target did not contradict any discourse ideas. The implicit inconsistent target sentences did present ideas that were inaccurate. However, because the readers had little familiarity with these ideas, the inaccuracies had little impact on them.

Correlations Between the Target Reading Times and Passage Familiarity Norms

The correlation between Singer et al.'s (1997) familiarity norms for the crucial ideas and the target reading times was computed in each experimental condition. Suppose that readers infer the relation between the clauses of the target sentence. Then, reading in the implicit consistent condition is likely to be faster for familiar than unfamiliar ideas, resulting in a negative correlation between reading times and norms. Conversely, a positive correlation would be expected in the implicit inconsistent condition. For the explicit conditions, the familiarity versus target reading time correlations would be tempered by the consistency or inconsistency between the explicit sentence and the target sentence.

The only familiarity versus target-sentence reading time correlations that approached significance in the two experiments were consistent with these predictions. In Experiment 1, there was a marginally significant negative correlation in implicit *consistent* condition, $r(22) = -.25, p = .12$. In Experiment 2, there was a positive correlation in the implicit *inconsistent* condition, $r(22) = .43, p = .04$. These results could come about only if the readers inferentially derived the crucial ideas from the clauses of the target sentences. None of the other correlations challenged the main hypotheses.

Agreement Rate

The proportions of "yes" replies to the crucial questions are shown in Table 11.3. For Experiment 1, ANOVA revealed a Consistency × Relation interaction, $F_1(1,66) = 28.5$, $MSE = 587$, $F_2(1,22) = 15.0$, $MSE = 200$. Tests of simple main effects revealed that explicit agreement exceeded implicit agreement for inconsistent items, $F_1(1,66) = 20.8$, $MSE = 459$, $F_2(1,22) = 31.7$, $MSE = 281$, but not for consistent items, $Fs \leq 2.56$.

There were main effects of all three variables of theoretical interest. Proportion agreement was 82% in the consistent condition and 57% in the

TABLE 11.3

Mean Proportion Agreement and Mean Correct Answer Times (in Milliseconds) for Experimental Questions in Experiments 1 and 2

			Proportion Agreement		*Answer Time*	
Experiment	*Set*	*Consistency*	*Explicit*	*Implicit*	*Explicit*	*Implicit*
Experiment 1	Singer	Consistent	.85	.85	2,979	3,240
		Inconsistent	.72	.55	3,332	3,459
	Noordman	Consistent	.82	.74	3,289	3,738
		Inconsistent	.70	.32	3,641	3,980
Experiment 2	Singer	Consistent	.95	.90	2,821	2,998
		Inconsistent	.71	.59	3,172	3,286
	Noordman	Consistent	.80	.78	3,106	3,428
		Inconsistent	.68	.53	3,389	3,460

inconsistent condition, $F_1(1,66) = 194.8$, $MSE = 462$, $F_2(1,22) = 46.1$, $MSE = 14,627$. There was a relation main effect, with agreement rates of 77% and 61% in the explicit and implicit conditions, respectively, $F_1(1,66) = 79.6$, $MSE = 456$, $F_2(1,22) = 32.3$, $MSE = 192$. Finally, rates of 75% and 64% were detected for the Singer and Noordman sets, respectively, although this difference was only marginally significant in the items-random ANOVA, $F_1(1,66) = 31.5$, $MSE = 538$, $F_2(1,22) = 3.45$, $MSE = 753$, $p = .08$. The Set × Relation effect was also significant, $F_1(1,66) = 14.8$, $MSE = 481$, $F_2(1,22) = 6.09$, $MSE = 1169$. This diagnosed a larger difference between explicit and implicit rates for the Noordman than the Singer materials.

In Experiment 2, analysis of the proportions of agreement with the experimental questions revealed a Relation × Consistency interaction, $F_1(1,61) = 18.1$, $MSE = 605$, $F_2(1,22) = 16.2$, $MSE = 109$. Tests of simple main effects revealed higher agreement rates for explicit than implicit items in the inconsistent condition, $F_1(1,61) = 46.4$, $MSE = 707$, $F_2(1,22) = 20.1$, $MSE = 286$; but not in the consistent condition, Fs ≤ 2.54.

Like in Experiment 1, all three main effects of theoretical interest were significant. The proportion agreements were greater for the explicit than the implicit condition, $F_1(1,61) = 42.3$, $MSE = 539$, $F_2(1,22) = 32.2$, $MSE = 542$; for consistent than inconsistent items, $F_1(1,61) = 214.7$, $MSE = 479$, $F_2(1,22) = 11.3$, $MSE = 378$; and for Singer passages than for Noordman passages, $F_1(1,61) = 36.8$, $MSE = 455$, $F_2(1,22) = 4.48$, $MSE = 661$.

Like the target reading times, the most important feature of the agreement rates was the Relation × Consistency interaction. In the inconsistent condition, explicit agreement rates exceeded the implicit rates. This is likely because the explicit inconsistent passages included one sentence (the explicit sentence) that supported the question and one (the target sentence) that contradicted it; whereas the implicit inconsistent passages presented only a single sentence (target) that contradicted the question.

In the consistent condition, in contrast, the agreement rate was very similar for explicit and implicit items. This indicates that the readers were as successful in deriving the critical fact from the target sentence alone as from the target plus explicit sentences. It is noteworthy that the explicit inconsistent acceptance rates were also substantially lower than those of both consistent conditions. These data profiles support the conclusion that the readers derived valuable information from the target sentences. Otherwise, explicit agreement rates would have exceeded the implicit ones, yielding only a main effect of relation.

Answer Times

The answer times for affirmative replies in both experiments are shown in Table 11.3. However, only the answer times in the consistent condition were subjected to ANOVA. Because the proportion agreement in the inconsistent condition ranged from 32% to 72%, many participants had one or more missing cells of affirmative answer times. In addition, in experimental conditions in which participants answer "yes" about 50% of the time, many of the replies contributing to that proportion are likely to be guesses. In such a circumstance, it is unlikely that a mean answer time reflects processing of a uniform sort. Therefore, the inconsistent condition answer times were not analyzed. It was expedient to apply a single ANOVA to the Consistent answer times of the two experiments.

In the participants-random ANOVA of the Consistent answer times, set and relation were within variables and experiment (Experiments 1 and 2) was a between variable. In the items-random ANOVA, experiment and relation were within variables, and set was a between variable.

The answer times were 404 ms longer for the Noordman than the Singer materials, $F_1(1,136) = 21.2$, $MSE = 407{,}729$, $F_2(1,44) = 7.49$, $MSE = 464{,}453$. Answer times were also 355 ms longer in the implicit condition than the explicit condition, $F_1(1,136) = 16.8$, $MSE = 407{,}729$, $F_2(1,136) = 45.3$, $MSE = 48{,}450$. In addition, there was a significant effect of experiment in the participants-random analysis only, reflecting faster responding in Experiment 2, $F_1(1,136) = 10.4$, $MSE = 896{,}352$. Finally, the Set × Relation interaction approached significance in the items-random ANOVA, $F_2(1,44) = 3.41$, $MSE = 48{,}450$, $p = .07$. This reflected the fact that implicit answer times exceeded explicit ones by 219 ms for the Singer passages and 385 ms for the Noordman passages. A test of simple main effects revealed a significant effect of relation even for the Singer passages, $F_1(1,136) = 6.36$, $MSE = 468,753$, $F_2(1,44) = 11.9$, $MSE = 48,450$.

The answer times present a complicated picture, relative to the main hypotheses. In this paradigm, equal explicit and implicit answer times favor the bridging inference position whereas greater implicit than explicit times is argued to deny inference processing. Like in Singer et al. (1997), the present consistent conditions yielded differences between the explicit and implicit answer times that were greater for the Noordman materials than the Singer materials. However, even for the Singer passages, implicit answer time exceeded explicit time. Singer et al. (Experiment 3) detected this pattern only when they encouraged slow readers to speed up, a constraint that we did not impose. The main distinction between the present method and that of Singer et al. was the frequent appearance (50% of all passages) of passage ideas inconsistent with the knowledge question that ensued. Whether the presence of inconsistencies in these materials can account for the relation effect detected for the consistent Singer sequences in this study cannot be resolved by the present experiments.

CONCLUSIONS

The results were very similar in the two experiments, and conformed in most respects with the assumption that readers construct and validate causal bridging inferences when they comprehend science texts of low familiarity. The main challenge to the validation analysis stemmed from the answer time data. We now address some central implications of these findings.

Reconciling Adjacent and Distal Text Ideas

It is notable that interposing two sentences between the explicit and target sentences did not prevent the readers from detecting inconsistencies between them (Experiment 2). In view of the complexity of the present materials, we raised the possibility that target reading time and question agreement rate might become dissociated in Experiment 2. To the contrary, both measures mirrored the patterns detected in Experiment 1. This outcome contradicts the possibility that readers reconcile the ideas derived from expository text only when those ideas co-occur in working memory. The intervening sentences of Experiment 2 should promote the deletion of the explicit sentence from working memory. However, there are some factors that *would* tend to preserve the explicit sentence in working memory. Those factors include the thematic status of that idea, the fact that it had no causal consequence (Fletcher & Bloom, 1988), and the conceptual overlap between the explicit sentence and the two intervening sentences (O'Brien & Albrecht, 1991). Direct tests of these alternatives remain as future endeavors in this project.

Material Sets

According to most of the dependent variables, the Noordman passages posed more difficulty than the Singer passages (see also Singer et al., 1997). It is noteworthy, therefore, that the data *patterns* were highly similar for these sets: In neither experiment did the Set × Relation × Consistency interaction approach significance for either reading time or agreement rate. It would be disagreeable to regularly detect different result patterns for two sets of materials, without being able to attribute the differences to some identifiable characteristic of the passages. In this regard, and notwithstanding the greater difficulty of the Noordman passages, Singer et al. (1997, Experiments 3 and 4) reported that several experimental manipulations exerted similar impacts on both sets of material.

According to the constructionist analysis of discourse comprehension (Graesser et al., 1994), coherence is preserved at multiple levels of discourse representation, and readers routinely retrieve the causes of text outcomes. The unfamiliar content of many expository texts might impede the inferential detection of connections among text ideas. As explained toward the outset, however, the complexity of expository text does not strictly prohibit fruitful inference processing (Kieras, 1985; Perrig & Kintsch, 1985). Consistent with the findings of Singer et al. (1997), the present results favor the alternative that readers of expository text engage in inference processes reminiscent of those that accompany narrative comprehension, even when the topic matter is unfamiliar. Of course, for expository texts on familiar themes, the ability to confidently validate tentative bridging inferences is likely to permit to the reader to construct a richer and more robust text situation model (McNamara & Kintsch, 1994).

ACKNOWLEDGMENTS

This research was supported by Grant OGP9800 to the first author from the Natural Sciences and Engineering Research Council of Canada. Experiment 1 was conducted by the second author in partial fulfillment of the requirements of the Honors B.A. degree at the University of Manitoba.

REFERENCES

Cheng, P. W., & Holyoak, K. J. (1985). Pragmatic reasoning schemas. *Cognitive Psychology, 17,* 391–416.

Fletcher, C. R., & Bloom, C. P. (1988). Causal reasoning in the comprehension of simple narrative texts. *Journal of Memory and Language, 27,* 235–244.

Goldman, S. (1997). Learning from text: Reflections on 20 years of research and suggestions for new directions of inquiry. *Discourse Processes, 23,* 357–398.

Graesser, A. C. (1981). *Prose comprehension beyond the word.* New York: Springer-Verlag.

Graesser, A. C., Singer, M., and Trabasso, T. (1994). Constructing inferences during narrative text comprehension. *Psychological Review, 101,* 371–395.

Grice, H. P. (1975). William James Lectures, Harvard University, 1967. Published in part as "Logic and conversation." In P. Cole & J. L. Morgan (Eds.), *Syntax and semantics: Vol. 3. Speech acts* (pp. 41–58). New York: Seminar Press.

Keenan, J. M., Baillet, S. D., & Brown, P. (1984). The effects of causal cohesion on comprehension and memory. *Journal of Verbal Learning and Verbal Behavior, 23,* 115–126.

Kieras, D. E. (1985). Thematic processes in the comprehension of technical prose. In B. Britton & J. Black (Eds.), *Understanding expository prose* (pp. 89–107). Hillsdale, NJ: Lawrence Erlbaum Associates.

McNamara, D. S., & Kintsch, W. (1994, November). *Learning from history texts: Effects of text coherence and background knowledge.* Paper presented at the meeting of the Psychonomic Society, St. Louis, MO.

Miller, J. R. (1985). A knowledge-based model of prose comprehension: Applications to expository texts. In B. Britton & J. Black (Eds.), *Understanding expository text* (pp. 199–226). Hillsdale, NJ: Lawrence Erlbaum Associates.

Myers, J. L., Shinjo, M., & Duffy, S. (1987). Degree of causal relatedness and memory. *Journal of Memory and Language, 26,* 453–465.

Noordman, L. G. M., Vonk, W., & Kempff, H. J. (1992). Causal inferences during the reading of expository texts. *Journal of Memory and language, 31,* 573–590.

O'Brien, E. J., & Albrecht, J. E. (1991). The role of context in accessing antecedents in text. *Journal of Experimental Psychology: Learning, Memory, and Cognition, 17,* 94–102.

Perrig, W., & Kintsch, W. (1985). Propositional and situational representations of text. *Journal of Memory and Language, 24,* 503–518.

Potts, G. R., Keenan, J. M., & Golding, J. M. (1988). Assessing the occurrence of elaborative inferences: Lexical decision versus naming. *Journal of Memory and Language, 27,* 399–415.

Singer, M. (1993). Causal bridging inferences: Validating consistent and inconsistent sequences. *Canadian Journal of Experimental Psychology, 47,* 340–359.

Singer, M., & Ferreira, F. (1983). Inferring consequences in story comprehension. *Journal of Verbal Learning and Verbal Behavior, 22,* 437–448.

Singer, M., & Halldorson, M. (1996). Constructing and validating motive bridging inferences. *Cognitive Psychology, 30,* 1–38.

Singer, M., Halldorson, M., Lear, J. C., & Andrusiak, P. (1992). Validation of causal bridging inferences in discourse understanding. *Journal of Memory and Language, 31,* 507–524.

Singer, M., Harkness, D., & Stewart, S. T. (1997). Constructing inferences in expository text comprehension. *Discourse Processes, 24,* 199–228.

Trabasso, T., & Sperry, L. L. (1985). Causal relatedness and importance of story events. *Journal of Memory and Language, 24,* 595–611.

Trabasso, T., van den Broek, P., & Suh, S. Y. (1989). Logical necessity and transitivity of causal relations in stories. *Discourse Processes, 12,* 1–25.

12

Comprehension of Mathematical Proofs

Charles R. Fletcher, Sarah Lucas, and Corinne M. Baron
University of Minnesota

Narrative comprehension has been an area of continuous inquiry in cognitive psychology since the late 1970s (e.g., Bower, Black, & Turner, 1979; Mandler & Johnson, 1977). During the last 15 years, much of this research has focused on the role that causality plays in the representation and processing of narratives. This is largely due to the influence of Trabasso and his colleagues (Trabasso & Sperry, 1985; Trabasso & van den Broek, 1985; van den Broek, 1988) who demonstrated empirically that:

1. Narrative states and events with many causal connections to the rest of the narrative are remembered better and perceived as more important than otherwise similar states and events.
2. States and events that lie along a causal path that connects the opening of a narrative to its final outcome are remembered better and perceived as more important than causal "dead ends."

These results led to the widely held belief that the mental representation of a narrative can be thought of as a complex network of states and events tied together by causal relations. The explicit goal of a great deal of subsequent research has been to develop process models that explain how these complex, richly interconnected representations are constructed within the limitations of the human information-processing system (e.g., Albrecht & O'Brien, 1993; Fletcher & Bloom, 1988; Graesser, Singer, & Trabasso, 1994; Kintsch, 1992; McKoon & Ratcliff, 1992; van den Broek, 1990).

In many respects, mathematical proofs are as different from narratives as any genre of discourse one could imagine. Where narratives are understood easily by almost all adults in our society, proofs are difficult or impossible for all but a few specialists. Where narratives are among the most common forms of discourse in our society, proofs are among the most esoteric. Where the ability to understand and compose narratives is learned, for the most part, without formal instruction, the ability to understand and compose proofs is not. Where the rules for composing a well-formed narrative are largely implicit, the rules for composing a well-formed proof are explicit. Where everyone enjoys a good narrative, the ability to enjoy a good proof is, unfortunately, quite rare. Finally, where narrative comprehension has received considerable attention from cognitive psychologists, the comprehension of proofs has been completely ignored.

Despite these differences, there is an apparent similarity between narratives and proofs. Consider the simple geometry proof in Fig. 12.1. It consists of some given information, a statement of what is to be proven, and an interconnected series of steps that lead, via enabling relations, from the given information to the conclusion of the proof. There is a strong analogy between this "solution path" through the proof and the "causal path" that connects the opening of a narrative to its final outcome. The strength of this analogy becomes even more obvious when we turn our attention to the criteria that are normally used to identify causal relations in narratives (e.g., Trabasso, van den Broek, & Suh, 1989). Typically, Event A is said to cause Event B if, in the circumstances of the text, Event B would not have occurred had Event A not occurred. This counterfactual criterion leads to a broad definition of causality that encompasses physical causality, psychological causality, motivation, and (as in mathematical proofs) enablement. When this criterion is applied to the proof in Fig. 12.1, it results in the "causal network" shown in Fig. 12.2. The only obvious difference between this network and those derived from narratives is the absence of causal dead ends (i.e., steps whose antecedents cannot be traced back to the given information or whose consequences do not lead to its ultimate conclusion).

The fact that we, as researchers, can construct a narrative-like causal representation of a proof raises the possibility that enabling relations might play a role in the mental representation of proofs similar to that played by causal relations in the mental representation of narratives. This, in turn, suggests that contemporary process models of narrative comprehension (e.g., Albrecht & O'Brien, 1993; Fletcher & Bloom, 1988; Graesser et al., 1994; Kintsch, 1992; McKoon & Ratcliff, 1992; van den Broek, 1990) could

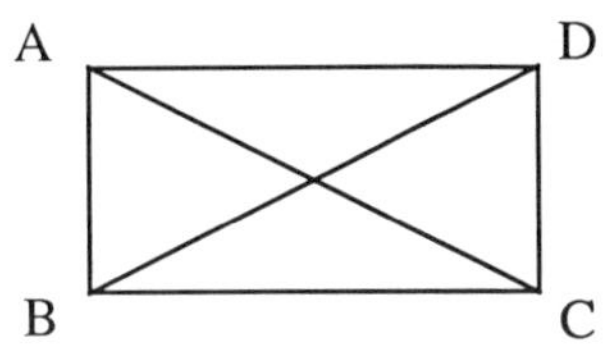

Given: ABCD is a parallelogram and AC $\cong$ DB

Prove: ABCD is a rectangle

Step	Justification
1. ABCD is a parallelogram	Given
2. AC $\cong$ DB	Given
3. AB $\cong$ DC	Step 1; Opposite sides of a parallelogram are $\cong$
4. BC $\cong$ BC	Step 1; Reflexive Property
5. ΔABC $\cong$ ΔDCB	Step 2; Step 3; Step 4; Side-Side-Side
6. $\angle$ ABC $\cong$ $\angle$ DCB	Step 5; Corresponding parts of $\cong$ Δs are $\cong$
7. $\angle$ ABC is supplementary to $\angle$ DCB	Step 1; In a parallelogram, consecutive angles are supplementary
8. $\angle$ ABC is a right angle	Step 6; Step 7; If angles are both $\cong$ and supplementary, they are right angles
9. $\therefore$ ABCD is a rectangle	Step 1; Step 8; If a parallelogram contains at least 1 right angle, it is a rectangle

FIG. 12.1. A simple geometry proof used in Experiments 1 and 2.

offer some insights into the cognitive processes involved in understanding proofs. To explore these possibilities, we conducted two experiments. In the first, we attempted to replicate with proofs one of the most robust results in the narrative comprehension literature (e.g., Trabasso & Sperry, 1985;

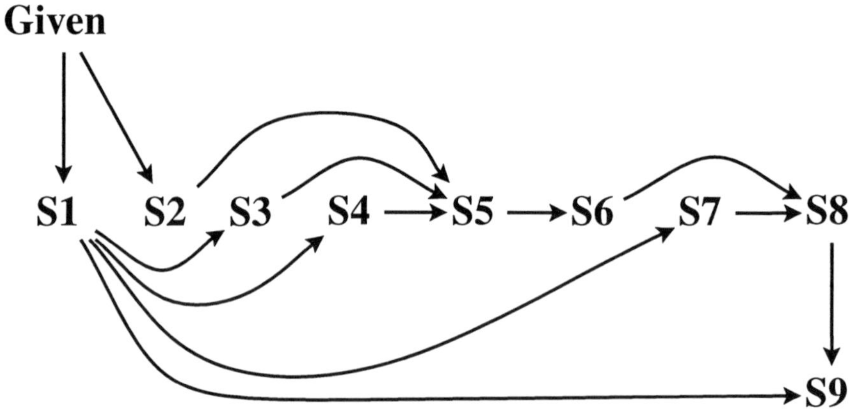

FIG. 12.2. A "causal diagram" based on the procedures of Trabasso, van den Broek, and Suh (1989) showing enabling relations between steps of the proof in Fig. 12.1. A link from Step A to Step B indicates that, in the context of the proof, Step B would not be possible without Step A.

van den Broek, 1988): the finding that events with many causal relations are perceived as more important than otherwise similar events with fewer causal connections. In the second experiment, we used verbal protocol analysis (Ericsson & Simon, 1984) to begin the process of comparing the cognitive processes involved in the comprehension of proofs to those involved in understanding narratives.

EXPERIMENT 1: THE MENTAL REPRESENTATION OF PROOFS

The goal of this experiment was to test the hypothesis that those steps in a proof that have many enabling connections to other steps in the proof (such as S5 in Fig. 12.2) will be perceived as more important than otherwise similar steps that have fewer enabling connections (such as S6). If this hypothesis proves false, we will know that the mental representation of proofs is fundamentally different from the mental representation of narratives. If this is the case, it is unlikely that process models of narrative comprehension will teach us anything about understanding proofs. But if the hypothesis is true, there is at least a chance that some aspects of these models will generalize to proofs. This experiment was closely modeled after the research of Trabasso and Sperry (1985).

Method

Participants. Participants in this study were 19 high-ability undergraduate students from colleges and universities throughout North America who were taking part in a federally funded summer research program for students in the social sciences at the University of Minnesota. As part of this program, all were enrolled in a Graduate Record Examination (GRE) preparation course and had recently completed a review of mathematics skills, including high school geometry. All participants were native speakers of English and received $5 in return for their participation.

Materials. The primary materials for this study were eight mathematical proofs. Seven were adapted from high school geometry texts and included diagrams (see Fig. 12.1). The final proof was a simple number theory proof (given that n is an odd integer, prove that n^2 is also odd) taken from a college text on reading and constructing proofs (Solow, 1990). It did not include a diagram. The number of steps in the eight proofs ranged from a low of 7 to a high of 14, with a mean of 9.6. The number of enabling relations between steps in the proofs ranged from 9 to 17 with a mean of 14.8. An additional geometry proof served as a warm-up item. All the proofs, together with a page of instructions, were gathered into test booklets. The order of the proofs, except for the warm-up item which was always presented first, was randomized within each booklet.

Procedure. Participants were instructed to work through their test booklets one proof at a time. They were asked to study each proof until they felt confident that they understood it. Then they were asked to rate the importance of each step on an 11-point scale. A rating of 1 indicated that a step was among the least important, that removing it would have no impact on the understandability of the proof as a whole. A rating of 11 indicated that a step was among the most important, that removing it would render the remaining proof incomprehensible. A rating of 6 indicated that a step was of average importance. Participants completed this study in groups of six or less in sessions that lasted up to 1½ hours. They were provided with a list of all the symbols used in the proofs together with their definitions. They were allowed to consult this list at any time.

Results and Discussion

The importance ratings were analyzed using procedures recommended by Lorch and Myers (1990). We began by performing 152 separate regression analyses, one for each combination of 19 participants and eight proofs. In each analysis, the number of enabling relations connecting each step to the rest of the proof was treated as the independent variable (see Fig. 12.1) and the participant's importance rating for that step was treated as the dependent variable. The resulting regression weight provides a measure of the relationship between number of enabling connections and perceived importance. The overall mean regression weight of .51 indicates that each enabling connection increases perceived importance by just over one half of a rating point. This value is reliably greater than zero when tested against subject variability, $t(18) = 2.48$, $p < .05$, but not when evaluated against item variability, $t(7) = .66$, $p > .10$. Given that only eight proofs were included in this research, the failure of the analysis by items to reach standard levels of statistical reliability is hardly surprising. Overall, these results closely mirror those of Trabasso and Sperry (1985) and suggest that the mental representation of proofs can reasonably be considered a narrative-like causal network as shown in Fig. 12.1.

EXPERIMENT 2: CONTROLLED PROCESSES IN COMPREHENSION

Given that there is at least some similarity between the mental representations of proofs and narratives, we now turn our attention to the processes responsible for constructing those representations. Intuition suggests that, for the vast majority of readers, understanding a proof is a more effortful, controlled process than understanding a narrative (Schneider & Shiffrin, 1977). This experiment had two related goals. The first was to confirm this intuition. The second was to explore its impact on the processing of proofs versus narratives. To achieve these goals, we asked participants to "think out loud" while reading both proofs and narratives. Previous research has shown that this procedure, when used properly (Ericsson & Simon, 1984), can reveal the nature of the controlled processes involved in both comprehension and problem-solving tasks (Ericsson & Simon, 1984; Fletcher, 1986; Newell & Simon, 1972; Olson, Duffy, & Mack, 1984; Suh & Trabasso, 1993). As such, it seems ideally suited for the task at hand. We expected to find that, relative to narratives, proofs would elicit more repetitions and paraphrases of the material being read, more comments about the relationship between the material being read and the rest of the discourse, and more comments about the comprehension process itself. This pattern of results would suggest that the process of constructing a causal representation like that depicted in Fig. 12.2 is a less automatic, more effortful process for proofs than for narratives.

Method

Participants. Participants in this study were 14 undergraduate students from the same population tested in Experiment 1, though no participant took part in both studies. All were native speakers of English and all received $5 in return for their participation.

Materials. The primary materials for this study were two of the proofs from Experiment 1 (including the proof in Fig. 12.1) and two simple narratives from Fletcher and Bloom (1988). The proofs included a mean of 8.5 steps with 10.5 enabling connections. The narratives averaged 21 states and events with 27 causal and enabling connections. One additional proof and one additional narrative from the same sources served as warm-up items. These materials were gathered into test booklets. Each booklet included the warm-up narrative, followed by the warm-up proof, followed by the remaining proofs and narratives in a random order. Within the booklets, each proof or narrative was presented across several pages. For the proofs, the first page included all the given information and the statement of what was to be proven. Each subsequent page retained the information from the previous page and added the next step in the proof. The new information on each page was always presented in boldface type. For the narratives, the first page included only the title of the narrative. Each subsequent page added the next sentence. Once again, the new information was always presented in boldface type.

Procedure. Participants completed this experiment one at a time in sessions lasting up to 1½ hours. Each session began with a series of warm-up exercises from Ericsson and Simon (1984) designed to familiarize participants with the process of thinking out loud. They were then asked to think aloud while reading the three proofs and three narratives. This was accomplished by working through the test booklets one page at a time. Participants were asked to read the new information on each page (in boldface type) out loud, then to verbalize any thoughts they had while reading and trying to understand it before proceeding to the next page. An experimenter was present during each session and prompted participants to "remember to think out loud" any time they read two consecutive steps or sentences without commenting on them. Each session was tape recorded and transcribed. To ensure a deep level of understanding, participants were told that a memory test would follow the presentation of the proofs and narratives. A free-recall test of the two experimental proofs and the two experimental narratives was administered at the end of the study, but the results were not analyzed. As in Experiment 1, participants were provided with a list of the symbols used in the proofs together with their definitions. They were allowed to consult this list at any time.

Results and Discussion

As a first step in analyzing the verbal reports, each protocol was parsed into statements corresponding to nonsubordinate clauses, each with its own main verb. The first reading of each new statement was excluded from further analysis. All remaining statements were then sorted into the categories shown in Table 12.1. These categories are based largely on a taxonomy of inferences proposed by van den Broek, Fletcher, and Risden (1993) and focus on the relationship between the participants' thoughts (as expressed by the statements in the verbal protocol) and the narrative event or proof step that they are attempting to understand (the focal step/event). A single judge sorted all of the statements. All subsequent analyses were based on her judgments. To access the reliability of these judgments, a second judge sorted the statements from five randomly selected protocols. The two judges assigned statements to the same category 87% of the time.

Figure 12.3 shows the relative frequency of each type of statement listed in Table 12.1 (except for the "other" category, which received no further consideration). A series of two-tailed t tests for matched samples was used to determine which categories occurred more frequently in proofs and which were more common in narratives. These tests all used error terms based on subject variability. Because only two proofs and two narratives were included in this study, no attempt was made to access the generalizability of the results across items.

Connecting inferences ($t(13) = -3.26, p < .01$), repetitions of the focal statement ($t(13) = -4.67, p < .001$), and metastatements ($t(13) = -2.21, p < .05$) were all reliably more likely to occur in proofs than in narratives. Thus,

TABLE 12.1

Definitions and Examples of the Categories Into Which Participants' Verbal Protocol Statements Were Sorted in Experiment 2

Category	*Definition*	*Example*
Connecting Inference	Comments on relations between the focal step/event and any other step/event in the discourse.	"If the diagonals are congruent, then you know that it's a rectangle."
Forward Elaboration	Comments that anticipate some future step/event in the discourse.	"This is going to have a lot to do with inner angles."
Backward Elaboration	Comments on a prior step/event not explicitly included in the discourse.	"Well, if you have a parallelogram, both sides of the quadrilateral are parallel."
Orthogonal Elaboration	Comments on a step/event not explicitly included in the discourse that neither precedes nor follows the focal step/event.	"They certainly look congruent in the figure."
Reinstatement	Any repetition or paraphrase of a step/event that explicitly preceded the focal step/event.	"We already know that the two lines in the middle of the parallelogram are congruent."
Association	Comments on personal knowledge or memories not directly relevant to understanding the discourse.	"Jeez, I hated these in high school!"
Repetition	Any repetition or paraphrase of the focal step/event after it is read the first time.	"AC is congruent to DB ... given ... AC is congruent to DB."
Metastatement	Any comment on the comprehension process itself.	"So, I don't really understand what they are doing here."
Other	Any comment not included in the categories above.	"Um ... let's see ..."

Note. All examples are from the proof in Fig. 12.1.

consistent with our expectations, participants devoted more conscious attention to the comprehension process itself, as well as to the focal statement and its relationship to the rest of the discourse while reading and understanding proofs. These findings support the claim that understanding a proof is a more effortful process than understanding a narrative.

The only categories that were reliably more likely to occur in narratives were forward elaborations ($t(13) = 4.04$, $p < .01$) and associations ($t(13) = 2.13$, $p < .10$). In other words, while reading narratives, participants were more likely to think ahead and more likely to be reminded of events in their own lives. The first of these findings is reminiscent of a result by Olson, Mack, and Duffy (1981), who found that readers are more likely to make forward inferences while reading narratives than they are while reading expository texts. In their research, as in ours, this difference is probably due

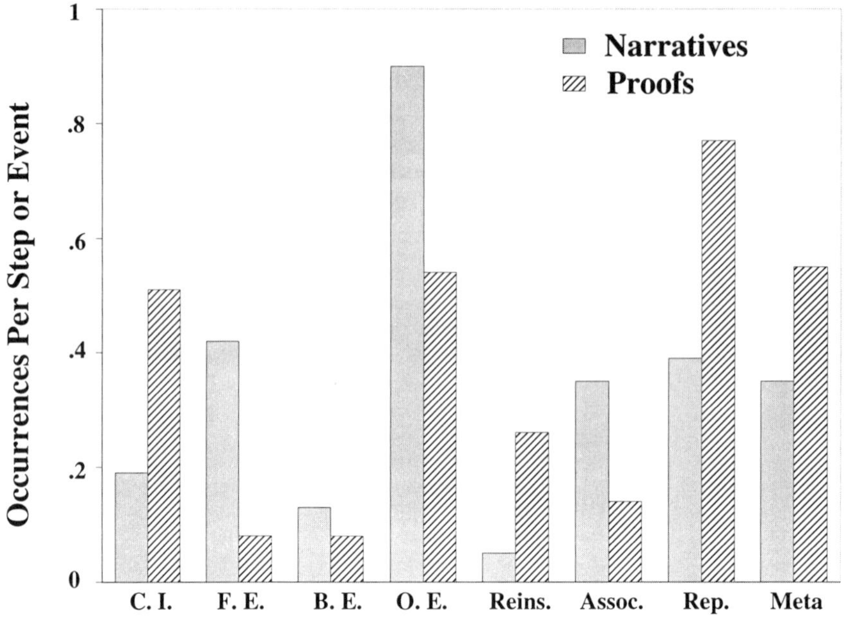

FIG. 12.3. Probability (per proof step or narrative event) of connective inferences (C. I.), forward elaborations (F. E.), backward elaborations (B. E.), orthogonal elaborations (O. E.), reinstatements (Reins.), associations (Assoc.), repetitions (Rep.), and metastatements (Meta) in the think-aloud protocols from Experiment 2 as a function of discourse type (proofs vs. narratives).

to the fact that readers have more cognitive resources available while reading narratives as opposed to more challenging genres of discourse. It remains to be seen whether the critical difference is genre per se, or level of difficulty. Figure 12.3 also suggests that orthogonal elaborations are more common in narratives, but this difference is due almost entirely to the influence of a single participant and is not statistically reliable.

SUMMARY AND FUTURE DIRECTIONS

The process of *constructing* mathematical proofs has long been considered a paradigm case of deductive reasoning. As such, it has been studied extensively by both psychologists (Koedinger & Anderson, 1990) and computer scientists (Nevins, 1975). As a result, it is somewhat surprising that the process of *understanding* proofs has been completely ignored by cognitive scientists. This lack of attention is especially puzzling when one considers the practical significance of the genre. The research reported here represents a first step toward filling this obvious void. It was motivated by the hope that contemporary models of narrative comprehension (e.g., Albrecht & O'Brien, 1993; Fletcher & Bloom, 1988; Graesser et al., 1994; Kintsch, 1992; McKoon & Ratcliff, 1992; van den Broek, 1990) can provide a useful starting point for exploring the processes of understanding and remembering proofs.

In Experiment 1, we found that enabling relations in a proof influence importance ratings in much the same manner that causal and enabling relations do in a narrative. We infer from this that the causal network representations popular in research on narrative comprehension also provide a reasonable approximation to the mental representation of a proof. This suggests that process models that attempt to explain how these representations are created might apply to proofs as well as to narratives. Experiment 2, however, provides a cautionary note. It confirms everyone's intuition that understanding a proof is a more effortful, controlled process than understanding a narrative. While reading proofs, our participants found it necessary to devote more conscious attention to understanding each step and its relations to the rest of the proof. They were also forced to devote extra attention to monitoring and controlling the comprehension process itself. As a result, they were less able to anticipate where the discourse was going or to notice relations between it and their own lives (e.g., "Will this proof help me figure out how much paint I need for the kitchen?"). These findings suggest that additional research is needed to determine where the similarities between these two genres end, and the differences begin. In our laboratory, we have turned our attention to the following questions: Does the number of enabling relations that a step has to the rest of the proof predict how well it will be recalled or how likely it is to be included in a summary of the proof (as in Trabasso & van den Broek, 1985)? Can our own process model of narrative comprehension and recall (Fletcher & Bloom, 1988; Fletcher, van den Broek, & Arthur, 1996) predict the flow of information through a proof reader's awareness? Can it predict which parts of a proof will be recalled best and which will be forgotten? It is our hope that the results reported here will encourage others to tackle similar questions.

In closing, we would like to point out that mathematical proofs offer comprehension researchers two major advantages over narratives. First, the ability to understand narratives is confounded with age. The ability to understand proofs is not. As a result, longitudinal and expert–novice studies of proof comprehension promise new insights into the acquisition of comprehension skills. Second, it is almost impossible to formally specify the background knowledge required to understand a narrative. This limits our ability to construct computer simulations of narrative comprehension (e.g., Fletcher et al., 1996; van den Broek, Risden, Fletcher, & Thurlow, 1996). In the domain of proofs, simulations of the construction process already exist (e.g., Koedinger & Anderson, 1990; Nevins, 1975). These include all of the background knowledge needed to understand, as well as construct, a proof.

ACKNOWLEDGMENTS

Sarah Lucas and Corinne Baron were supported by National Science Foundation Grant 9600021 to the Cognitive Sciences Center at the University of Minnesota. We are grateful to Paul van den Broek for his comments on an earlier draft of this chapter and to all the members of the Minnesota Text Comprehension Research Group for their many helpful and encouraging suggestions.

REFERENCES

Albrecht, J. E., & O'Brien, E. J. (1993). Updating a mental model: Maintaining both local and global coherence. *Journal of Experimental Psychology: Learning, Memory, and Cognition, 19,* 1061–1070.

Bower, G. H., Black, J. B., & Turner, T. J. (1979). Scripts in memory for text. *Cognitive Psychology, 11,* 177–220.

Ericsson, K. A., & Simon, H. A. (1984). *Protocol analysis: Verbal reports as data.* Cambridge, MA: MIT Press.

Fletcher, C. R. (1986). Strategies for the allocation of short-term memory during comprehension. *Journal of Memory and Language, 25,* 43–58.

Fletcher, C. R., & Bloom, C. P. (1988). Causal reasoning in the comprehension of simple narrative texts. *Journal of Memory and Language, 27,* 235–244.

Fletcher, C. R., van den Broek, P., & Arthur, E. (1996). A model of narrative comprehension and recall. In B. K. Britton & A. C. Graesser (Eds.), *Models of understanding text* (pp. 141–163). Hillsdale, NJ: Lawrence Erlbaum Associates.

Graesser, A. C., Singer, M., & Trabasso, T. (1994). Constructing inferences during narrative text comprehension. *Psychological Review, 101,* 371–95.

Kintsch, W. (1992). How readers construct situation models for stories: The role of syntactic cues and causal inferences. In A. F. Healy, S. Kosslyn, & R. M. Shiffrin (Eds.), *Essays in honor of William K. Estes* (pp. 261–278). Hillsdale, NJ: Lawrence Erlbaum Associates.

Koedinger, K. R., & Anderson, J. R. (1990). Abstract planning and perceptual chunks: Elements of expertise in geometry. *Cognitive Science, 14,* 511–550.

Lorch, R. F., & Myers, J. L. (1990). Regression analysis of repeated measures data in cognitive research. *Journal of Experimental Psychology: Learning, Memory, and Cognition, 16,* 149–157.

Mandler, J. M., & Johnson, N. S. (1977). Remembrance of things parsed: Story structure and recall. *Cognitive Psychology, 9,* 111–151.

McKoon, G., & Ratcliff, R. (1992). Inference during reading. *Psychological Review, 99,* 440–466.

Nevins, A. J. (1975). Plane geometry theorem proving using forward chaining. *Artificial Intelligence, 6,* 1–23.

Newell, A., & Simon, H. A. (1972). *Human problem solving.* Englewood Cliffs, NJ: Prentice-Hall.

Olson, G. M., Duffy, S. A., & Mack, R. L. (1984). Thinking-out-loud as a method for studying real-time comprehension processes. In D. Kieras & M. Just (Eds.), *New methods in the study of immediate processes in comprehension* (pp. 253–286). Hillsdale, NJ: Lawrence Erlbaum Associates.

Olson, G. M., Mack, R. L., & Duffy, S. A. (1981). Cognitive aspects of genre. *Poetics, 10,* 281–315.

Schneider, W., & Shiffrin, R. M. (1977). Controlled and automatic human information processing: I. Detection, search, and attention. *Psychological Review, 84,* 1–66.

Solow, D. (1990). *How to read and do proofs: An introduction to mathematical thought processes.* New York: Wiley.

Suh, S., & Trabasso, T. (1993). Inferences during reading: Converging evidence from discourse analysis, talk-aloud protocols, and recognition priming. *Journal of Memory and Language, 32,* 279–300.

Trabasso, T., & Sperry, L. L. (1985). Causal relatedness and importance of story events. *Journal of Memory and Language, 24,* 595–611.

Trabasso, T., & van den Broek, P. (1985). Causal thinking and the representation of narrative events. *Journal of Memory and Language, 24,* 612–630.

Trabasso, T., van den Broek, P., & Suh, S. Y. (1989). Logical necessity and transitivity of causal relations in stories. *Discourse Processes, 12,* 1–25.

van den Broek, P. (1988). The effects of causal relations and hierarchical position on the importance of story statements. *Journal of Memory and Language, 27,* 1–22.

van den Broek, P. (1990). The causal inference maker: Towards a process model of inference generation in text comprehension. In D. Balota, G. Flores d'Arcais, & K. Rayner (Eds.), *Comprehension processes in reading* (pp. 423–445). Hillsdale, NJ: Lawrence Erlbaum Associates.

van den Broek, P., Fletcher, C. R., & Risden, K. (1993). Investigations of inferential processes in reading: A theoretical and methodological integration. *Discourse Processes, 16,* 169–180.

van den Broek, P., Risden, K., Fletcher, C. R., & Thurlow, R. (1996). A "landscape" view of reading: Fluctuating patterns of activation and the construction of a stable memory representation. In B. K. Britton & A. C. Graesser (Eds.), *Models of understanding text* (pp. 165–187). Mahwah, NJ: Lawrence Erlbaum Associates.

13

Content Integration and Source Separation in Learning From Multiple Texts

M. Anne Britt
University of Pittsburgh
Charles A. Perfetti
University of Pittsburgh
Rebecca Sandak
University of Pittsburgh
Jean-François Rouet
University of Poitiers, France

In recent years, we have been studying how students learn and reason from history texts, with the emphasis on the plural—multiple texts. Wc began with a simple assumption about the form of the typical history textbooks read by students, namely that they were essentially narratives. This implied that students' understanding of history texts would typically include a representation of temporal-causal event sequences. Work by Tom Trabasso and his colleagues on the causal analysis of narratives provided a fine starting point for our goal of describing this temporal-causal component in history stories. Building on early story-understanding research (Mandler & Johnson, 1977; Omanson, 1982; Rumelhart, 1975; Stein & Glenn, 1979; Thorndyke, 1977), Trabasso and colleagues proposed that causal connections play a critical role in establishing meaning and coherence in a story. By providing a detailed and explicit model of causal analysis and spelling out tests for necessary and sufficient conditions, Trabasso and colleagues were able to predict what readers recall (Trabasso & van den Broek, 1985), include in a summary (Trabasso & van den Broek, 1985; van den Broek & Trabasso, 1986), and judge as important (Trabasso & Sperry, 1985; Trabasso & van den Broek, 1985). Although this type of causal analysis clearly captured an essential part of narrative understanding, an analysis of the use of multiple texts required more than this.

Our goal here is to explain the course our work on multiple texts has taken, given its initial foundation on the simple idea of history as narrative and understanding as taking account of temporal-causal event structures. First we describe our initial work to develop temporal-causal models that reflect the more complex demands of long multiple texts and summarize what we learned in one study. Then we describe a later phase of our work that required the development of an additional level of text representation to capture what readers understand about the source as well as the content of texts. In the last section, we briefly summarize the results of a recent study that tests possible models of how readers represent source information.

ADAPTING CAUSAL MODELS TO MULTIPLE TEXTS

From a student reader's point of view, a salient component of learning history is learning events and the causal-temporal connections among these events. Indeed, although historical scholarship goes well beyond the event narrative, it is fair to say that reconstructing the causes that explain events is a critical part of much written history. To mirror the narrative component of a typical text, a reader's representation must capture these temporal-causal relations, making history learning similar to general story understanding. However, for reasons we give later, it became clear that one cannot simply use existing analyses (e.g., Trabasso & van den Broek, 1985) to capture these representations. Some modifications were required.

Our first modification was to increase the grain size. Understanding the complex events related through actual history texts required us to use experimental texts that were much longer and more complex than the narrative texts previously studied. We intended to allow students to learn from texts that were between 2,000 and 13,000 words in length rather than the few hundred words typical in story-understanding research. A clause-by-clause analysis was impractical, whereas event units were both more practical and more appropriate for the level of understanding we wanted to capture. Increasing the grain size from the clause to the level of the event solved the problem. Each event unit can be elaborated either in a single clause or in several paragraphs; each higher level event, in effect, is a substory, more or less elaborated depending on the knowledge and purpose of the author. Our assumption here is that causal structures are valid for macroevents (e.g., the June 6, 1944 Normandy invasion CAUSED the Western Front military defeat of Nazi Germany) just as they are for microevents (the ringing of the alarm clock CAUSED General Montgomery to wake up that morning).

Our second modification was to develop a shared template to capture common events described by different sources. Our assumption was, and is, that learning historical narratives requires the reader to construct a model of the situation from many sources and to integrate information from these sources rather than constructing a separate model of each text. To capture the kind of information that would be represented in an integrated model, we needed an abstracted representation of events and their surrounding elements that were shared among the various texts. To accomplish this, we read dozens of sources about the Panama Canal story and then constructed a single common template of the events and causal-temporal structures that comprise the typical Panama Canal story. The abstract causal-temporal event template developed for this story is shown in Fig. 13.1.

Using this template, we were able to describe the story contained in a selected set of texts on the Panama Canal and study a small group of students who learned from them (Perfetti, Britt, & Georgi, 1995). Over a 4-week period, the students read a lengthy (approximately 15 pages) scholarly or popular text describing the U.S. acquisition of a canal in Panama at the turn of the century. After reading at home, students came into the lab to write both a short and a long summary of the events as they understood them so far. They then proceeded to answer a series of comprehension questions and reasoning probes. Applying our causal-temporal event template to the information provided by students in their summaries and in answers to our questions, we obtained three general results. First, during their initial reading, students rapidly and extensively learned *core events* (i.e., those that were central to the story), and a limited number of the supporting details for these events. Second, over subsequent readings, students gradually learned the *noncore events,* that is, events that were not necessary to a coherent telling of this story. Students also continued to learn supporting details for both core and noncore events with each reading. Third, although they learned more of the complete story (core and noncore events elaborated with supporting details) over time, the composition of the students' sum-

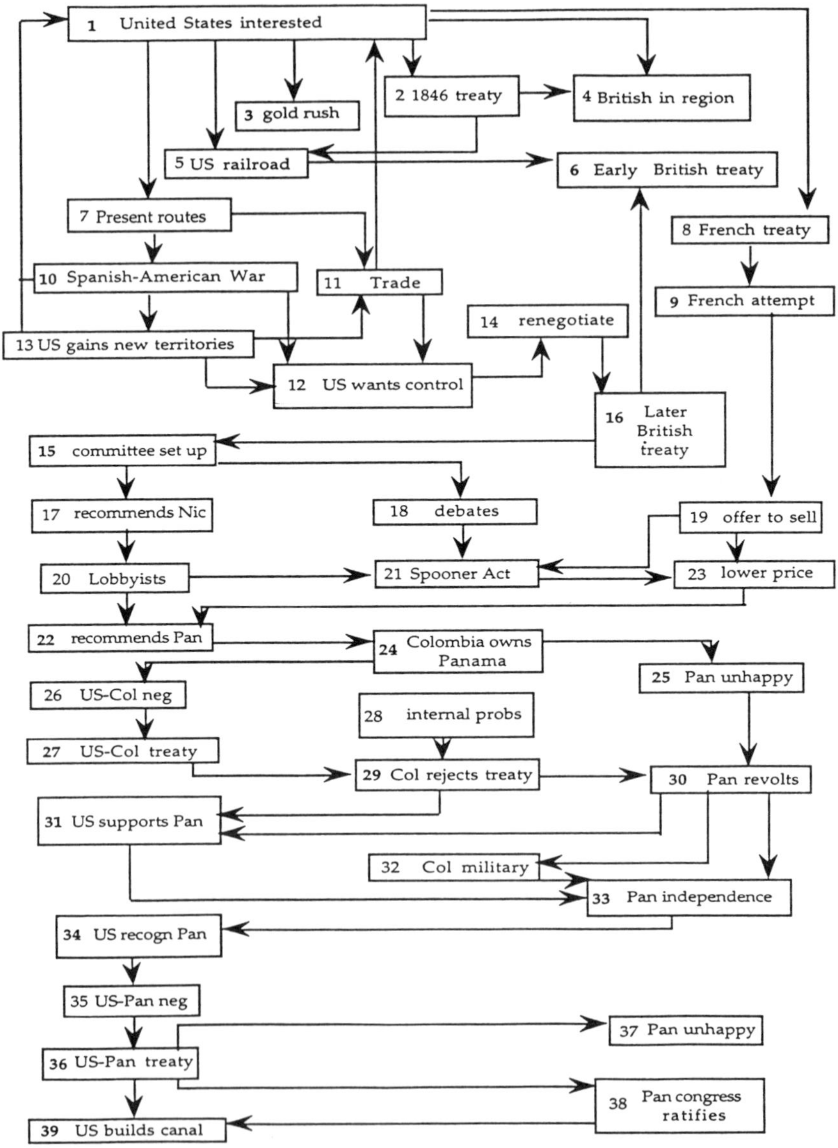

FIG. 13.1. Causal-temporal event template of the history of the acquisition of a U.S. canal in Panama. From Perfetti, Britt, and Georgi (1995). Copyright 1995 by Lawrence Erlbaum Associates. Adapted by Permission.

maries remained rather stable. Their summaries were predominantly a synopsis of the core events and the causal-temporal connections of these events.

These two developments (larger grain size and a common template for shared events) enabled us to represent temporal-causal narrative structure across several texts of substantial size and to effectively describe the time course of learning across readings in the Perfetti et al. (1995) study. These developments also enabled us to represent the integrated accounts that the students provided in their summaries and answers to questions.

Further analysis of the student interviews revealed several occasions on which students accurately attributed information to a specific author. This need to represent the distinct contribution of a given text to the story is apparent in many real multiple-text learning situations, and it is particularly salient in history. History texts comprise multiple, and frequently contradictory, causal models. A representation of the texts must include not only what is shared among them, but also what is not shared and what is discrepant. When several possible causal connections exist between events, coherence of the representation cannot be estimated solely by the causal connectedness of a single text. For instance, two authors may propose incompatible causes for a single event. By allowing events or a series of events to be associated with the document's source, readers can maintain overall coherence in a mental model while still allowing for otherwise incoherent causal connections when integrating information across text. The reader can remember that "according to X, A caused B"; whereas "according to Y, C caused B." Thus a comprehensive model of multiple-text representation must include the potential for source information, and it becomes an empirical question as to the conditions that promote the encoding of source information.

This interesting combination of integration and separation of material from different sources led us to consider the range of possibilities for how to represent the kind of complex integrated-yet-separated representation that learners may develop from multiple texts. Do learners keep track of source information while they are learning the shared story? Surely, we thought, some learning of source information is likely, but it is unlikely that every piece of information is tagged for its source. Theoretically, the question is how the causal-temporal event model might include source tagging for its events. Our armchair considerations of the possibilities led us to four classes of models. Before considering these models, we first examine studies that appear to address these issues of integration and separation.

Evidence of Integration

Wineburg (1991) had historians and advanced high school history students read several documents and think aloud during problem solving. He found that historians employed a "corroboration" heuristic, which involves comparing important information against what one knows from other sources before integrating the information. Wineburg showed that experts, but not high school students, integrate information derived from multiple sources and they do this integration while reading.

Further evidence of integration comes from a series of studies by Wiley and Voss (1996; Voss & Wiley, 1997). College students read about the Irish Potato famine from eight separate sources or from a textbooklike chapter that weaved a story out of the same information. Wiley and Voss observed that students who read separate sources used more connectors in their essays and judged possible inferences from the texts better, whereas the single source lead to better verbatim recognition of information. These studies imply integration of information across separate sources leads to greater knowledge transforming rather then knowledge telling.

Evidence of Separation and Source Tagging

Several other studies of multiple document reading in history suggest that at least some source tagging is conducted as the reader goes through the texts. Wineburg (1991), in the study described previously, found an additional strategy employed by the expert historians. They used a "sourcing" heuristic, the encoding and evaluating of source information before reading the contents of a document. Because historians learn about the source of a document before reading, they have the information necessary to make connections between source and content during reading. Recall that experts also corroborate while reading, which requires them to use this source tagging to reactivate previous authors' versions of the events and causes. Both the sourcing and the corroboration heuristics are precisely the type of processing that would facilitate the construction and use of a representation that includes source tagging.

One of our previous studies provides further evidence that readers' model of the information contained in multiple texts is augmented with source tagging. Rouet, Britt, Mason, and Perfetti (1996) gave 24 college students several problem statements about the U.S. acquisition of a canal in Panama (e.g., To what extent was the U.S. military intervention in the Panamanian revolution justified?). For each problem statement, they read short excerpts from different types of documents (e.g., textbook, historian essays, participant accounts). Students wrote essays to answer each problem and then ranked each document in the set as to its trustworthiness and its usefulness. Two findings support the hypothesis that students had encoded a link between the author and its content. First, 56% of student essays contained at least one explicit reference to the source of a statement. Most of these (91%) were, in fact, correct. Second, in analyzing the justification of their trustworthiness and usefulness ranks, we found that many had identifiable origins (e.g., this document was useful because it "cites precedent of guarding the railroad, but not interdicting Colombian troops"). Of the 110 justifications that were specific enough to trace the origins to a single document, 88% mentioned at least one correct statement of content linked to a source. Furthermore, 92% of these justifications were accurate. Even without specific instructions, students did source tag at least some content.

These results demonstrate that readers can take note of specific sources among a set of related documents under conidtions that promote problem solving. The question of the generality and spontanteity of sourcing remains to be determined. In a talk-aloud procedure, Kushner (1996) asked both undergraduate and graduate college students to verbalize their thoughts during the reading of history texts without any expectation of essay writing. Although the instructions did not refer to source information, all readers made comments about the source—as opposed to the content—of what they were reading, in one form another—references to the author directly, or to document characteristics. Interestingly, graduate students made more comments about sources than undergraduates. Thus, Kushner's results suggest, at least in these explicit verbalizing conditions, both that readers attend to source information spontaneously during reading and that individual differences in this attending are related to education level.

The research thus far suggests that a reader's representation of multiple texts is partially integrated and at least sometimes includes some links between source information and its content. Beyond this, there seems to be little to constrain the theoretical possibilities. To give full recognition to the range of these possibilities, the next section outlines four models of how tagging may be represented. Then in the last section, we describe an experiment (Britt, Sandak, Perfetti, & Rouet, 1998) that tests the predictions of the four models.

MODELS REPRESENTING SITUATIONS AND SOURCES

The four possible models of how readers might represent information learned from two or more related texts vary along a dimension of integration from completely separate and nonoverlapping situation models of each text to a single highly integrated situation model of the events. The models also vary in the degree of source tagging from no tagging to complete tagging of all content.

Separate Representation Model

The *separate representation* model, illustrated in Fig. 13.2, assumes that readers form a unique representation of each text without connections among them. Although individual propositions are not tagged, the entire unique representation may be tagged if author information is prominently provided. Information from each text is not connected to information from any other text (i.e., not integrated). Consider the two texts shown in Fig. 13.2, which are excerpts from the Britt et al. (1998) materials that are described in the next section. The fictitious authors, Clark and LaCosta, are represented by shaded nodes in the diagram, which we refer to as *document nodes.* Document nodes represent information known about the source and the document itself and are connected to content through bidirectional links. The top part of the diagram represents the events (rectangles) and causal connections (solid lines) from Clark's text, whereas the bottom portion represents the events and causal connections from LaCosta's text. The defining feature of this model is that there are two distinct situation models, one for each text, and they are completely independent of each other. Here the reader fails to make a connection between the events that overlap between each author's version of the story (i.e., fails to integrate).

Under most circumstances, this model would be less than optimal because prior knowledge (i.e., the situation model of the first author's text) was not accessed during the reading of the second text. There are conditions, however, when readers may be more likely to form a separate representation model of multiple texts. For example, if the source of the text is distinct and elaborated (i.e., who the author is and what type of document it is), the reader may be more able to keep the representations clearly distinct. Furthermore, if the task is to learn author X's version of the story, a reader may inhibit the activation of prior knowledge during the encoding process,

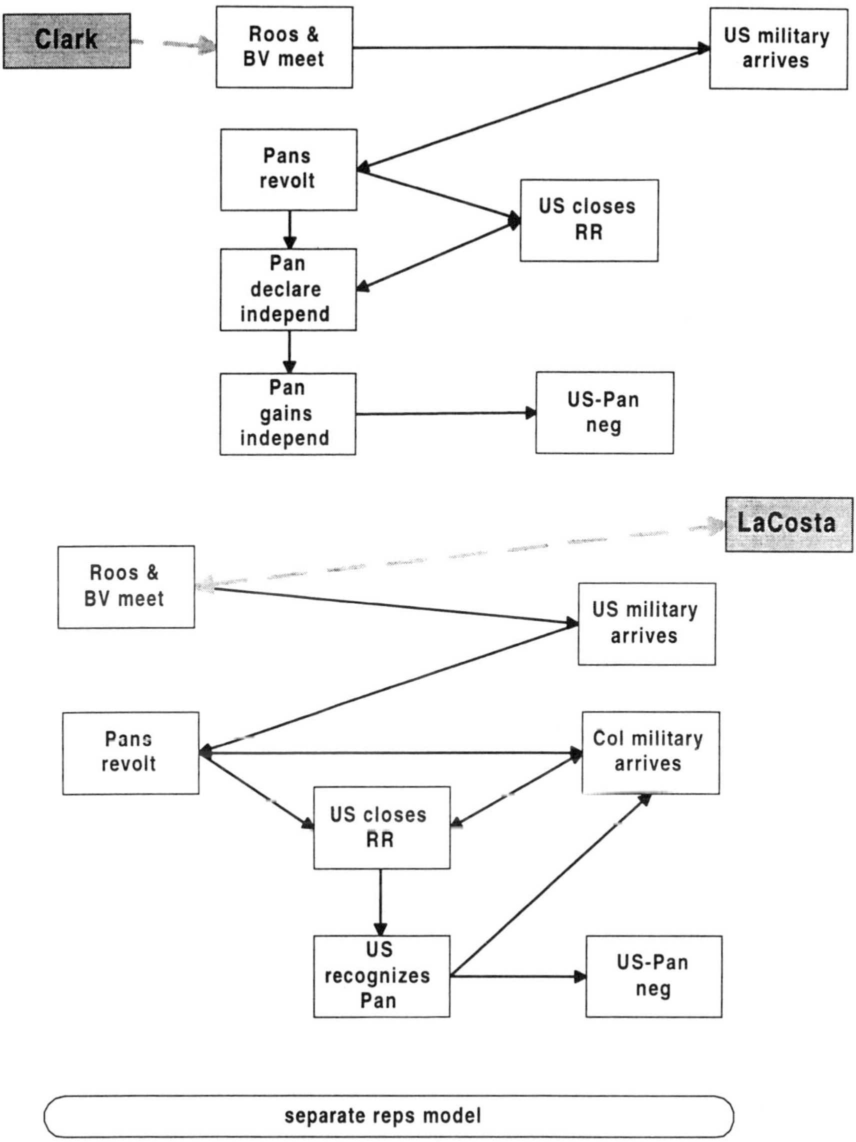

FIG. 13.2. Separate representation model representing a segment of text from two authors (Clark and LaCosta). Representation of situation (boxes and solid arrows) and intertext links (dotted lines) to a document node (shaded rectangles).

thus resulting in two unintegrated situation models. Anything that hinders integration will more likely lead to this type of representation. For example, if the texts do not highly overlap much in content then they may not be sufficiently related to trigger the activation of the other representation as the new situation model is created. The same might be expected if too much time has elapsed between readings. Either of these factors would result in a lessened ability to integrate events during the processing of the second text.

Mush Model

The *mush model* is essentially the opposite of the separate representation model. As shown in Fig. 13.3, the reader forms a combined representation that integrates information learned from all relevant texts. The essential element of this model is that there is no tagging or marking for where the information came from. All information is integrated completely, without attention to the source of the information. In the illustration, note that there

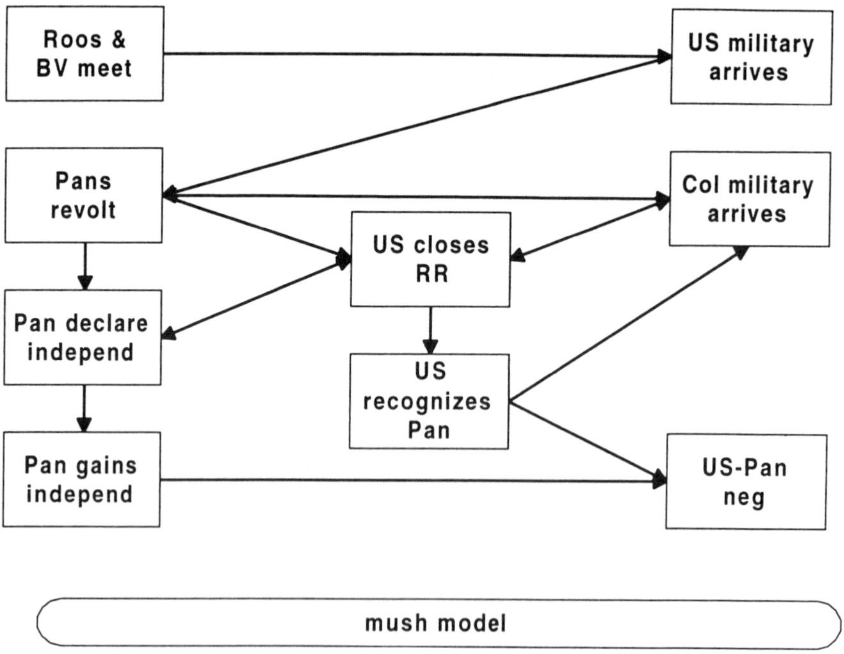

FIG. 13.3. Mush model representing a segment of text from two authors (Clark and La-Costa). Authors are not represented nor are the links between the author and the situation.

are no document tags on the events to reflect the model's assumption that author information is no longer available in memory.

Like the separate representations model, this representation is less than optimal. Whereas the first model lacks integration, the mush model lacks source marking. Readers may be more likely to form the mush model when the texts' sources are either never mentioned or not elaborated. A directed focus on simply learning the story, as opposed to solving some problem that involved text contradictions, may also increase the probability that the reader will form an integrated representation. Anything that encourages integration will increase the probability that this type of representation will be formed. This includes a high degree of overlapping events mentioned among texts and a short time span between reading of the multiple texts. We suspect that this form of representation is most common in students learning in school settings where they often see their task as one of learning about the topic, and because all sources are presented with relatively equal status, there is no reason to pay attention to where information comes from. The mush model, although promoting a situational understanding, leaves the learner short on source understanding and makes it difficult to verify information and evaluate it.

Tag-All Model

Whereas the mush model might be said to characterize minimalist learning of the kind that professors complain about in their undergraduates, the *tag-all model* is one for the scholar. As shown in Fig. 13.4, each and every event is tagged for its source. The shaded document nodes indicate the source information for that document and the dotted lines link the document node to the situation model derived from that author's text. There is also a link between document nodes to indicate the relationship between the two sources (e.g., support vs. oppose, agrees with vs. disagrees with, gives evidence for vs. gives evidence against, etc.).

The problem with this model is the high processing demands it implies. Each piece of information must be tagged for its source and integrated into a situation model. The only way out of this problem would appear to be a dubious assumption that links between the document node and each element in the situation model can be generated automatically, or at least with no cost. Such an assumption, we think, is unwarranted for typical student readers, but may be appropriate for content and author experts. Scholars who already have a rich knowledge base that includes the texts of a certain

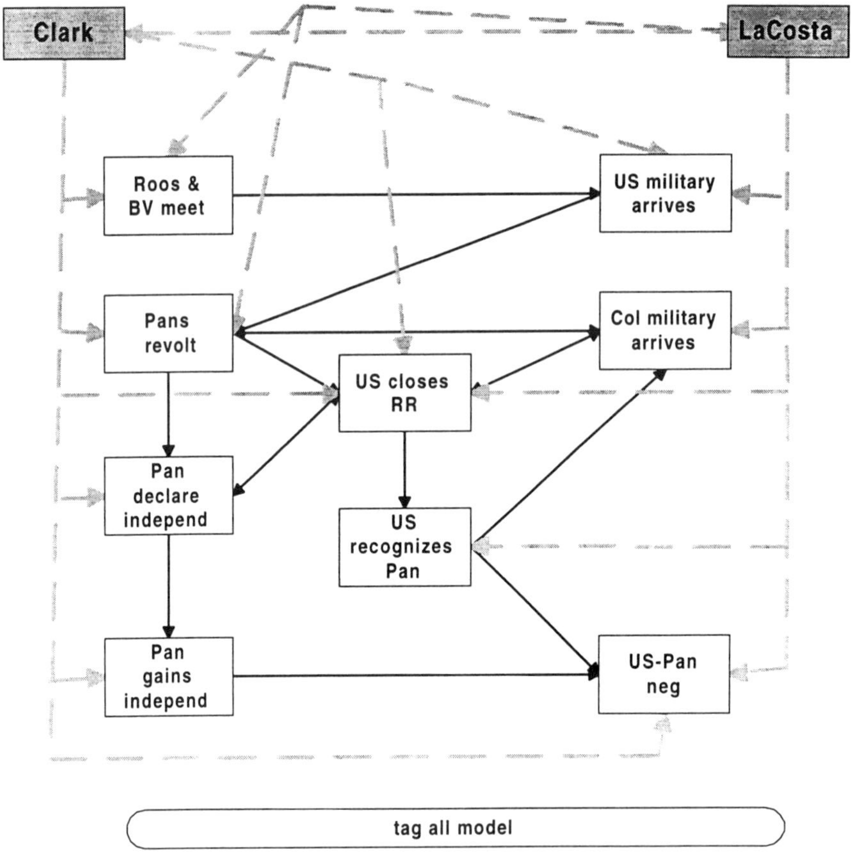

FIG. 13.4. Tag-all model representing a segment of text from two authors (Clark and LaCosta). Representation of situation (boxes and solid arrows) and intertext links (dotted lines) to a document node (shaded rectangles).

writer may come to "automatically" connect content and source for this author under some conditions. For a less expert reader, the effort may be too great.

Documents' Model

The model we proposed as most typical of a good reader's model of multiple-text learning in history is the *documents' model* (Perfetti, Rouet, & Britt, 1999), which has two interconnected components: a *situation model* and an *intertext model.* The intertext model is an added level of representation that includes an interconnection of links among document nodes and links from these document nodes to one or more nodes in the situation model. The intertext model for our two experimental texts is shown in Fig. 13.5 with the dotted lines representing the links between document nodes and relevant situation model nodes. The documents' model is a highly integrated situation model of the events learned, with only the most important events (i.e., core events) tagged for the source, as represented in the intertext model. Comparing Figs. 13.4 and 13.5, the difference between the tag-all model and the documents' model should be immediately apparent. Figure 13.5 shows only limited marking of source information as indicated by the links from the shaded Document Nodes.

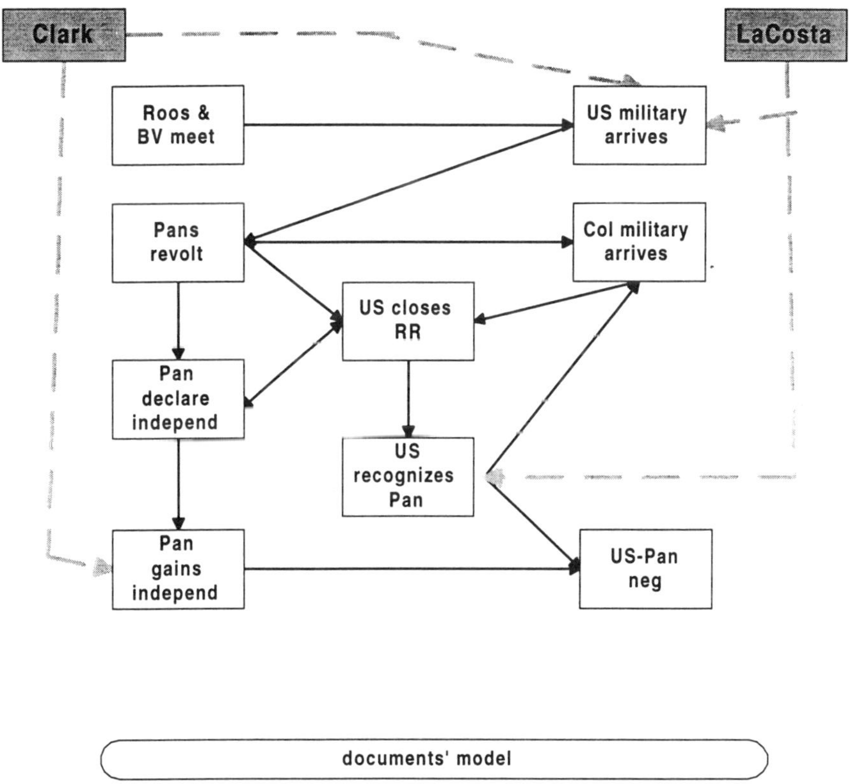

FIG. 13.5. Documents' model representing a segment of text from two authors (Clark and LaCosta). Representation of situation (boxes and solid arrows) and intertext links (dotted lines) to a document node (shaded rectangles).

Document nodes, which allow marking of events in the situation model, when fully elaborated contain information about the document's *source* (including information about the author, setting, and form of the document), the author's *rhetorical goals* (including the author's intents and perceived audience), and a synopsis of the *content* of the document. The document node for a particular text will be more or less elaborated depending on the quality of information provided to a reader about the source and also the reader's task (e.g., learning from more than one text written by the same author). This information can be used by the reader to connect a piece of information to its source (e.g., "according to author X ...") and to state the rhetorical relationships among texts (e.g., "based on Y, author X claims that ...", "document X contradicts document Y ...", etc.).

The conditions that promote or discourage source tagging need to be empirically determined. There is nothing intrinsic to the documents' model framework that constrains information tagging. Rather it is a model that allows, in principle, complete tagging or no tagging, with parameters to be established by considerations outside the model; for example, data of the kind we summarize next or a theory of text "importance" in which causal information (Trabasso & van den Broek, 1985) is more likely to be tagged would provide constraints on tagging. It is also possible that what gets tagged might be predicted by a computational text-comprehension model such as the construction-integration model (Kintsch & Welsch, 1991), the capacity-constrained construction integration model (Goldman, Varma, & Coté, 1996), or the landscape model (van den Broek, Risden, Fletcher, & Thurlow, 1996) that allows importance to arise from basic text-processing parameters.

Our initial approach was to reason that the core events of a situation are marked for source more than noncore events. This reflects our belief that readers who are motivated to attend to sources during reading manage the task complexity by marking what is important rather than what is not. We further thought that initially mentioned events may benefit from sourcing, whereas interference might affect tagging of later read events. The important constraint guiding these arguments is the assumption that processing limitations require nonexperts to tag only a small subset of both events and details.

Readers who are knowledgeable about a text's source will be more likely to form a documents' model. Similarly, it will be easier to form a document node when a text's source is salient and well elaborated. Even with unelaborated sources a reader could still code the two sources as "the first guy" and "the other guy," but a more semantically elaborated source would enable a reader to build a model of the knowledge and motives of the author, thus supporting inferences about likely claims and evidence mentioned by the author. A reader would also be more likely to form a documents' model when the task is to learn the story while keeping texts separate and when the texts provide conflicting accounts of partially overlapping events. Characteristics of the reader, such as their knowledge and experience using source information, will influence whether a reader attends to source information during reading as well.

EVIDENCE FOR AN INTEGRATED SOURCE-TAGGED REPRESENTATION

A recent study (Britt et al, 1998) examined some of the factors that might affect whether readers mark sources and, in so doing, tested the relative plausibility of the models just described. First, the study tested the assumption that sourcing would be associated more with core events than with noncore events. Second, it examined two task conditions that we believe affect how likely readers are to build integrated versus source-separated models. Specifically, we assumed that instructions to attend to sources would encourage source tagging compared with more typical comprehension instructions. And we thought that information that discredited one of the sources might discourage the building of an integrated model that lacked source tagging.

The study presented 80 students with well-controlled pairs of texts that, although sharing some events and details, differed in others and expressed different interpretations. Both texts related a narrative version of the Panama Canal story wrapped in either a pro-U.S. or an anti-U.S. argument. We selected 16 complex event propositions (e.g., Panama gained its independence through a rebellion) and two details supporting each complex event (e.g., the rebellion occurred with no bloodshed and it was brief). To test the hypothesis that sourcing accuracy varies with the centrality of the event in the causal-temporal event model, half of the 16 complex events were core events (those events that are critically important to telling a summary of story) and half were noncore events (those events that elaborate the story but can be omitted in an adequate summary of the story). To test readers' knowledge of the source of a particular proposition, we manipulated its *uniqueness,* that is, whether the proposition was contained in both texts (shared) or only one (unique). To control for the number of times a proposition was read—which would otherwise be two times for shared information but just once for unique information—items that were unique (i.e., to be mentioned only by one author) were mentioned exactly twice by that author and shared items were mentioned exactly once by each author. Each event had two critical details that were also mentioned precisely once or twice depending on whether the event was shared or unique. Whether within or across texts, the two mentions of a target proposition were paraphrases of the abstract proposition. Finally, in order to eliminate differences in memorability due to the specific propositions themselves, the study created two versions of each story. The two versions controlled the assignment of shared versus unique information, so that what was shared in one version was unique in the other. The result of this procedure was to unconfound shared versus unique with specific information within a text.

Groups of subjects participated in four different conditions that manipulated the two factors we expected to matter for sourcing: reading instructions (*sourcing* or *comprehension*) and discrediting information applied to one author (early, late, or never). In the sourcing instructions given to three of the groups, learners were told that they would be given two texts that would present different perspectives on the events and later they would have to indicate which author (if any) mentioned a particular event or detail. In comprehension instructions, given to only one group, nothing was said about sources; rather participants were told to learn the story and to be prepared to be tested on particular events and details. Finally, we also manipulated when, or if, participants were given information discrediting one of the authors (i.e., they were told that the first author was not very credible and should be believed only if the information was given by both authors). Two of the three sourcing instruction groups and the comprehension instruction group were given discrediting information, which always applied to the first text after it was read. The discrediting information was provided early (discredit the first text immediately after reading it), late (discredit the first text after both texts had been read but before writing the essay), and never (no discrediting information). Thus, the four groups were: sourcing instruction with discrediting early, sourcing instruction with discrediting late, sourcing instruction with discrediting never, comprehension instruction with discrediting late.

The procedure can be summarized as follows. All students read the first text; at that point, one group (sourcing-early discredit) was told that the first text was not trustworthy. This condition, we thought, should discourage readers from integrating information across the two texts. Next, for all groups, the first text was removed and the second text was read. After reading the second text, students were given a 1-minute distractor task to reduce text recency effects. Then they were given a list of propositions (sentences that paraphrased information that was in one or both texts or information that actually was not in the texts) and asked which author or authors, if any, mentioned the information. The test consisted of 24 items from each of these types: first author only, second author only, both authors, neither author (i.e., foils—half that contradicted facts mentioned and half that were true but unstated in the presented texts). Following this multiple-choice recognition test, two groups (sourcing-late and comprehension-late) were given discrediting information. Finally, all students were asked to write a short essay summarizing the events in Panama by including only information that was trustworthy. The final two tasks were the Nelson–Denny test of reading ability and a modified version of the verbal reasoning section of the Law School Admission Test. These were given to gain information about individual differences in our student sample that could be correlated with our performance measures.

Predictions

According to the separate representation model, one would expect readers to be very accurate at identifying an author, given that the content was represented. This would require only a search of the two individual situation models to find the information. Although these sequential searches may require high effort, they should lead to accurate identification. Because there is no integration, this model would predict that unique items would be better identified than shared items. We predict this because two mentions within a text would make the representation for that proposition stronger in the individual representation, thus more likely to be recalled. Without integration, there would be no strengthening of overlapping propositions during the second reading. Because shared items are weaker within a text, and there is no increased salience due to reinstatement of that proposition during the second reading, we predict that any advantage comes from more accurately sourcing unique items over shared items. Additionally, the unique items may have an advantage over shared items if the search is terminated after finding the proposition in the first author's situation model.

In contrast, the mush model predicts that participants will not be very accurate in determining who said what. Although, in this model, readers should do well at recognizing that a proposition was contained in what they read (content information), they should have trouble indicating which author mentioned it (source information) because propositions are not marked for source. Essentially, they should be at chance.

The tag-all model, like the separate representation model, should result in extremely accurate performance because everything is tagged. Although there may be better content recognition for core events, there should not be better source recognition for core events. Core and noncore events should be equally well recognized because both are tagged. Shared items should be sourced better than unique items because the model is integrated. While reading the second text, any propositions already in memory will be marked for the new source. This may increase the likelihood that the reader will at that time mark the proposition for the original text. Thus, both important and unimportant items will both be tagged at this time.

The documents' model should lead to accurate performance primarily on the core event items, because they are tagged. Noncore event items are more likely to not be tagged during reading. As with the tag-all model, the joint sources of shared items should be identified better than the single source of unique items because the reader is building an integrated model. During reading of the second text, information that is recognized as already represented becomes tagged, in effect strengthening source information even for noncore events. Readers can tag shared items that were not previously tagged (e.g., noncore events), thus leaving a predicted difference in the accuracy of sourcing shared items. This leads to the prediction of a specific interaction between centrality and uniqueness: Core events should be sourced better than noncore events, but only for unique items.

We turn now to consider selectively the results that are most important for distinguishing among these four models, emphasizing how they differ in representing source information.

Recognition Results

Content Recognition. Students were able to adequately distinguish presented items from not presented items, suggesting that they encoded and remembered content information from the texts. For items actually from the text, they had a hit rate of 82%; for foil items (neither author) they had a correct rejection rate of 70%, resulting in an overall average d′ of 1.64.

Source Recognition. Given that students were able to distinguish presented from nonpresented content, how accurately were they able to recognize the author of the content? To address this question, we looked at responses that identified one or the other or both sources. Overall students were able to identify the actual author at a better than chance rate, 56% of responses compared to a 33% chance level. This suggests that a model that assumes full integration and no representation of sources (the mush model) cannot be correct. Students were able to identify the source of content they recognized, not perfectly, but rather well relative to the chance level implied by the mush model.

Table 13.1 gives the average conditional percent correct recognition for centrality (core vs. noncore) and uniqueness (unique vs. shared). There was a significant effect of centrality with core events identified more accurately than noncore events. Only the documents' model predicts that more important items are more likely to be marked. There was also a significant effect of uniqueness, with shared items identified more accurately than unique items (63% > 48%). Recall that for every item, the number of mentions was not confounded with whether information was shared or unique. Thus, this result seems to reflect a genuine advantage for information that is shared between the texts over information that is unique to one text. This result argues against the separate representation model, which appears to predict either an advantage for unique information or no difference, depending on what assumptions are made about text procedures. The uniqueness effect, however, does support the documents' model and the tag-all model. Both models predict that during the integration process evoked by the second text, information encountered in the second text that was also contained in the first text will be tagged for both sources. Finally, there was a significant interaction of centrality by uniqueness, specifically, a smaller effect of centrality for shared than unique items. This is exclusively consistent with the documents' model, which predicts that core events are more likely to be tagged, as found with the unique items. For shared information, the noncore events should be noticed during the integration processes and should be tagged at that point. Thus, shared items should not have an appreciable benefit for centrality, as we found.

In summary, only the documents' model is consistent with all the recognition results. Models that allow only integration are ruled out by the

TABLE 13.1

Mean Conditional Percent Correct Recognizing the Author Identity by Centrality and Uniqueness

	Uniqueness	
Centrality	*Unique*	*Shared*
Core Events	54	64
Noncore Events	42	61

high level of source accuracy. Models that involve sourcing based on full tagging or on separate representations are ruled out by the selective sourcing of core information.

Trustworthy Essays

The recognition data show that students built a representation that included markings for the sources of some of the information. Whether students can use their source-marked representation in a meaningful way, such as writing an essay they consider trustworthy, is another question. By requesting students to write essays that were trustworthy, we are asking them to use information that they believed to be correct. By telling three groups of students that one of their sources was not an expert after all, but a politician who wrote to influence public opinion, we were suggesting that they perhaps should not use information from this discredited source in their essay. However, because the discrediting information always came after students had read the source to be discredited, the options for separating the two sources were limited. For complete success, students would have needed to tag all the information from the second text and then used that exclusively in writing their essay. That option was available to the early discredit group, who received discrediting information prior to reading the second text; however the late discredit group had already read the second text prior to receiving discrediting information. Their only chance to write a selective essay, that is, drawing only from the second and not from the first text, was to have already tagged the sources during reading.

In analyzing the student essays, Britt et al. (1998) made judgments of the source of each proposition contained in a student essay. For our purposes, the most important data are those based on conditional categorizations that took account of the students' judgment of that proposition in the recognition test. For each proposition in the student essay, Britt et al. asked how the student sourced it in the recognition test. For example, for a proposition written by both authors, the proposition was classified as unique if the student indicated on the recognition test that only one of the authors had mentioned it. Then each proposition in the essay could be classified according to the subject's judgment of its source: the discredited text (Clark), the nondiscredited text (LaCosta), or both. The basic result, shown in Table 13.2, is that students who were given discrediting information about Clark wrote essays that contained more information drawn from LaCosta, the trustworthy source, than did the group who received no discrediting infor-

TABLE 13.2

Mean Number of Essay Statements, Classified by Recognition Results, for Each Condition and Text

	Text	
Condition	Clark (Discredited)	LaCosta (Credited)
Sourcing instruction with discrediting early	.35	1.20
Sourcing instruction with discrediting late	.45	1.05
Sourcing instruction with discrediting never	.75	.60
Comprehension instruction with discrediting late	.45	1.00

mation. It did not matter whether the discrediting information came early or late or whether comprehension or sourcing instructions were provided.

INTEGRATION AND SEPARATION: A SUMMARY OF PROGRESS

We have argued that in learning from multiple texts a reader can achieve two goals that at first suggest incompatibility—developing an integrated representation of situations while keeping the source of the information separate. This is accomplished by the reader's acquisition of a documents' model, with the selective tagging of source information that allows integrated situations and partially separated sources.

In the present development of the documents' model, two components interact to affect the overall situational coherence a reader might establish from reading multiple texts. The intertext model, which includes document nodes to represent source information as well as links among sources and between sources and their content, proves especially useful in cases where several possible causal connections exist between the events described by different authors. When readers are confronted with texts that only partially overlap in their situational connections, they cannot compute a coherent representation on the basis of causal connectedness alone. When two authors propose incompatible causes for a single event, the intertext model allows these incompatible situations to be integrated and separated at the same time by source tagging each substory (or event sequence). Source tagging explains how readers maintain overall coherence in a mental model when causal connections are incoherent. The source-content connections override the event-event causal connections.

In general support for the framework of the documents' model is the study by Britt et al. (1998), which found that students were frequently able to identify which source had contained a given proposition. Information shared between the texts was identified more often than information that was unique to one text, supporting the assumption that information is integrated during reading. Supporting the selectivity assumption, and the corollary assumptions that information important to the narrative structure is privilege for selectivity, was the observation that students identified core events more than noncore events. The documents' model predicts both integration and clear but limited source tagging. Finally, this study found that the effect of centrality (core events) was stronger for information that was unique to one text than for information that was shared. We suggest that whereas core events are marked during the first reading, the unmarked events not central to the story are activated during the reading of the second text and marked during a corroboration process. Finally, the study also found that students were able to use this source-tagged structure when writing an essay. They could selectively use information from a source that had not been discredited and avoid information from a source that had been discredited.

The finding that sourcing is controlled by how central information is to a story establishes one element of the documents' model. Determining additional influences on the creation of source-content links requires further research. One clear influence should come from cross-referencing. An author who cites another source and links that source to the situation immediately provides a source link for the student's document model. This, in effect, gives for free what must be earned in learning situations such as the one we explored, where each source ignores all other sources. Other likely influences center on the task. Whereas our study allowed what was important to emerge from the texts themselves, importance can be established explicitly in the task. In the Panama Canal case, students might be given documents on Panama and asked to read them so that they can evaluate what President Roosevelt thought about the situation in Panama. This should strongly encourage tagging Roosevelt information that is relevant to the task. Nor must instructions be so explicit to be effective. Rouet et al. (1996), in requiring students to read documents in order to come to an informed opinion on a controversy, would have implicitly encouraged students to read from the perspective of the controversy, thus making controversial events more important to source.

The results from the Britt et al. (1998) experiment, in conjunction with prior results, show that readers can represent source information and its connection to content. Clearly, this ability is one that can be cultivated, and indeed it is not a routine ability even for college students. Both reading ability and verbal reasoning ability influenced the accuracy of this source tagging in Britt et al. If students are expected to learn from more than a single textbook excerpt, it helps if they are skilled at reading arguments and considering sources. The problem is that many students are not skilled in this way, and seem to have little experience prior to advanced college courses in such text study.

We are addressing this problem in a project that teaches high school students to attend to document sources using a computer-based learning environment. Some research has suggested that high school students do little spontaneous sourcing and corroboration (Wineburg, 1991), and students' implicit document knowledge is expressed only under some conditions (Britt, Rouet, & Perfetti, 1996; Rouet et al., 1996). The computer environment we have developed, called the Sourcer's Apprentice, presents students with a historical controversy and a set of excerpts from several different types of documents. The students identify different source characteristics (e.g., who the author is, when the document was written, etc.), answer questions about the content of the excerpts, and then write a short argumentative essay about the controversy. Additionally, Sourcer's Apprentice provides explicit tutoring about source characteristics and corroboration strategies and it provides directed practice exercises. We believe that intelligent use of texts must include paying attention to source information during reading and that training in the use and evaluation of sources will help students deal more effectively in multiple-text situations.

CONCLUSION

In a sense, we have come full circle in studying history text learning. We began by assuming and then verifying that learning from multiple history texts was, for the nonexpert, a question of learning a story. And story learning was about learning temporal-causal event structures. When we turned to developing a more elaborate account of the kinds of information readers could acquire, the idea of sources linked to situations was a natural extension. Now, in examining what students actually learn about sources when they read may take us back to the narrative itself. The students in the Britt et al. (1998) study appeared to selectively source documents in accord with the documents' representation model (Perfetti et al., 1999). Core information appears to be privileged in sourcing, just as it is in narrative history learning (Perfetti et al., 1995). Students appear most likely to mark sources for elements that are central to the story.

For purposes of history learning—at both the narrative and the source levels—we have renewed reason to appreciate the need for a rigorous accounting of causal structures of the kind developed by Trabasso and colleagues. In terms of the documents' model, whether we examine the situations model component or the intertext model component, we find evidence that learners take account of what is "important" in a particular way: Events that are core to the temporal-casual structure of the story are learned and if different sources create differences in this structure, that fact also will be noticed. Additional questions about what these circumstances are and how they control the learning situation will presumably take us far from temporal-causal analysis, but at first look, there is reason to believe that the narrative structure not only provides the form of what the reader learns about the situation, but also provides the platform from which to erect links to sources.

REFERENCES

Britt, M. A., Rouet, J.-F., & Perfetti, C. A. (1996). Using hypertext to study and reason about historical evidence. In J.-F. Rouet, J. J. Levonen, A. P. Dillon, & R. J. Spiro (Eds.), *Hypertext and cognition* (pp. 43–72). Mahwah, NJ: Lawrence Erlbaum Associates.

Britt, M. A., Sandak, R. L., Perfetti, C. A. & Rouet, J.-F. (July, 1998). *Content integration and source separation in learning from multiple texts.* Paper presented at the Annual meeting of the Society for Text and Discourse. Madison, WI.

Goldman, S. R., Varma, S., & Coté, N. (1996). Extending capacity-constrained construction integration: Toward "smarter" and flexible models of text comprehension. In B. K. Britton & A. C. Grasser (Eds.), *Models of understanding text* (pp.73–113). Mahwah, NJ: Lawrence Erlbaum Associates.

Kintsch, W., & Welsch, D. M. (1991). The construction-integration model: A framework for studying memory for text. In W. E. Hockley & S. Lewandowsky (Eds.), *Relating theory and data: Essays on human memory* (pp. 367–385). Hillsdale, NJ: Lawrence Erlbaum Associates.

Kushner, J. M. (1996) *The role of author information in reading.* Unpublished doctoral dissertation, University of Pittsburgh, Pittsburgh, PA.

Mandler, J. M., & Johnson, N. S. (1977). Remembrance of things parsed: Story structure and recall. *Cognitive Psychology, 9,* 111–151.

Omanson, R. C. (1982). The relation between centrality and story category variation. *Journal of Verbal Learning and Verbal Behavior, 21,* 326–337.

Perfetti, C. A., Britt, M. A., & Georgi, M. C. (1995). *Text-based learning and reasoning: Studies in history.* Hillsdale, NJ: Lawrence Erlbaum Associates.

Perfetti, C. A., Rouet, J.-F., & Britt, M. A. (1999). Toward a theory of documents representation. In H. van Oostendorp & S. R. Goldman (Eds.), *The construction of mental representations during reading.* Mahwah, NJ: Lawrence Erlbaum Associates.

Rouet, J.-F., Britt, M. A., Mason, R. A., & Perfetti, C. A. (1996). Using multiple sources of evidence to reason about history. *Journal of Educational Psychology, 88,* 478–493.

Rumelhart, D. E. (1975). Notes on a schema for stories. In D. G. Bobrow & A. Collins (Eds.), *Representation and understanding: Studies in cognitive science* (pp. 211–236). New York: Academic Press.

Stein, N. L., & Glenn, C. G. (1979). An analysis of story comprehension in elementary school children. In R. O. Freedle (Ed.), *New directions in discourse processing* (pp. 53–120). Hillsdale, NJ: Lawrence Erlbaum Associates.

Thorndyke, P. W. (1977). Cognitive structures in comprehension and memory of narrative discourse. *Cognitive Psychology, 9,* 77–110.

Trabasso, T., & Sperry, L. (1985). Causal relatedness and importance of story events. *Journal of Memory and Language, 24,* 595–611.

Trabasso, T., & van den Broek, P. (1985). Causal thinking and the representation of narrative events. *Journal of Memory and Language, 24,* 612–630.

van den Broek, P., & Trabasso, T. (1986). Causal network versus goal hierarchies in summarizing text. *Discourse Processes, 9,* 1–15.

van den Broek, P., Risden, K., Fletcher, C. R., & Thurlow, R. (1996). A "landscape" view of reading: Fluctuating patterns of activation and the construction of a stable memory representation. In B. K. Britton & A. C. Graesser (Eds.), *Models of understanding text* (pp. 165–187). Mahwah, NJ: Lawrence Erlbaum Associates.

Voss, J. F., & Wiley, J. (1997). Developing understanding while writing essays in history. *International Journal of Educational Research, 27,* 255–265.

Wiley, J., & Voss, J. F. (1996) The effects of playing historian on learning in history. *Applied Cognitive Psychology, 10,* 63–72.

Wineburg, S. S. (1991). Historical problem solving: A study of the cognitive processes used in the evaluation of documentary and pictorial evidence. *Journal of Educational Psychology, 83,* 73–87.

14

On the Use of Narrative as Argument

James F. Voss
University of Pittsburgh
Jennifer Wiley
Washington State University at Vancouver
Rebecca Sandak
University of Pittsburgh

THE NARRATIVE

Narrative is an important genre of discourse. It consists of the telling of a story, with the story having a temporal sequence, characters, actions, and events. In psychological research, the narrative has received considerable attention with a wide range of issues being investigated. Such issues include children's understanding of narrative (e.g., Stein & Glenn, 1979), causal reasoning (e.g., Trabasso, van den Broek, & Suh, 1989), inference generation (e.g, Graesser, Singer, & Trabasso, 1994), and models of comprehension and understanding (e.g., Kintsch, 1998).

The Narrative: General Characteristics

White (1987), and similarly McGuire (1990), distinguished three types of narrative, namely, historical, fictional, and fiction that is made to look like history. Historical narratives consist of the construction of accounts of past events, typically by historians, that are based on and derived from available documents. Generally, readers assume such accounts are accurate, given the constraints of the evidence and the beliefs and cultural background of the historian, as well as perhaps the historian's proclivity toward exaggeration. The classic volume by Thucydides (1972) on the Peloponnesian Wars is an example of such an account. Essentially the same documentation may be used, however, to construct two different accounts of the same set of events. Cronon (1992) demonstrated this fact by comparing two accounts of the history of the Great Plains. Thus, historical narratives provide information about the past and often the present, and the background of the narrator is frequently regarded as important (Mink, 1987).

The fictional narrative consists of a "story" in which the reader generally assumes that there are no "real" people and that the events portrayed did not really happen. The story is assumed to be a function of the author's imagination, as conditioned by the author's experience. The "truth" of fiction thus lies in fiction's portrayal of the human condition, the beliefs, emotions, thoughts, and reactions of people.

The idea of making fiction look like history refers to the use of historical figures or situations that are employed to infer a type of historical legitimacy to the story and actions of the characters. What Sturken (1997) has termed the "docudramas" of Oliver Stone serve as an example. Sturken was critical of movies such as *J.F.K.* because of a preoccupation with conspiracy theory and paranoia. Such stories are different than historical fiction because they are not written to be evaluated by the criterion of historical evidence even though some components of the story may embrace historical "fact."

Narratives are used for many purposes. In addition to conveying information and contributing to storytelling, they are used to integrate pieces of information. This is apparent in historical narratives, in which the historian will weave events and actions into a historical account, thereby providing a sense of chronological continuity. But as Hayden White (1987) commented, narrative discourse "endows them [real events] with illusory coherence" (p. ix). Narratives are used to tell stories of myth and folklore, and to transmit cultural values from generation to generation. Indeed, in a social-cultural context, Foucault (1969/1972) argued that people in power use narratives to maintain power, and those out of power have alternative narratives that are usually suppressed. In this context, Epstein (1996) showed that African American children frequently learn different historical narratives at home than at school and Wertsch and Rozin (in press) discussed the "official" and "unofficial" narratives of history found in Russia.

Of importance to the present chapter is that some writers have discussed a parallel of history and courtroom procedure (Collingwood, 1946; Wineburg, 1991). Whereas Collingwood spoke of the role of narrativity in both, such as when the prosecuting and defense attorneys describe their respective cases in narrative terms, Wineburg described the attorney as acting more like a historian whereas a reader of history was acting more like a juror. The prosecuting attorney, for example, analyzes the evidence, the testimony, and the presented documents in detail while the juror listens.

With respect to the present chapter, two aspects of the historical narrative, as related to the courtroom, are of particular interest. One is that the prosecuting and defense attorneys, much like the historians referred to by Cronon (1992), typically construct alternative narratives using the same evidence. In one case the attorney concludes the defendant is guilty; in the other, the conclusion is not guilty. Second, the goal, or at least an important goal, of the historians as well as of courtroom attorneys, is rhetorical; that is, the goal is persuasion. The historian generally wants to convince other historians and/or readers of the case being made, and the courtroom attorney wants to convince the jury. The idea that persuasion is being accomplished via use of a narrative brings us to the next issue.

NARRATIVE AS ARGUMENT

The use of the narrative by the historian or attorney to persuade essentially means that the narrative is being used as an argument component. Indeed, Aristotle (cited in McGuire, 1990) indicated that in the rhetorical context there are two types of argument, the enthymeme and the narrative.

Our position is that when the prosecuting attorney provides his or her statement, it has the general form of an enthymeme, that is, a claim supported by a reason. Thus, the statement "Capital punishment should be abolished because capital punishment kills innocent people" is an enthymeme. However, in the courtroom case, the reason supporting the claim is the narrative. Thus, the prosecuting attorney may state, "X is guilty because... ," and what follows the "because" is a narrative. Such a narrative usually tells how and why the crime happened and the defendant's role in it.

The prosecuting attorney's narrative, which is the supporting reason for the claim that "X, the defendant, is guilty," is assumed to have two components. One is the "facts" of the case, the testimony and the exhibits provided in the case. The second component is the narrative. The narrative is used to weave the facts of the case into a story that demonstrates the guilt of the defendant and hopefully to persuade the jury. Thus, the narrative supports the claim of guilty to form the argument.

But how are arguments evaluated? One criterion for evaluating arguments is that the reason supporting the claim must be acceptable. The second criterion is that the reason must provide support for the claim. This criterion is on occasion called relevance. In addition, a third criterion sometimes stated is that counterarguments need to be taken into account (Angell, 1964). In a real courtroom case, this criterion would suggest that, for example, when evaluating the prosecutor's case for the defendant's guilt, the defense attorney's argument should be taken into account. However, in the present study, we limited ourselves to the prosecution's side of the case. We therefore are concerned only with the first two evaluation criteria. None of these criteria is binary; reasons can be acceptable to different degrees, as can the support provided by the reason for the claim. Thus, reason acceptability and reason support for the claim could each be weak, moderate, or strong. For present purposes, it is assumed that the "jurors" base their evaluation on both criteria, that is, acceptance of the reason (narrative) and the support the reason (narrative) provides to the claim "X is guilty."

In the present study, the facts of the case were held constant across different narrative conditions; that is, they occur in the narrative in each narrative condition. Given this, it would be expected that if the prosecutor provided a well-constructed narrative, and "jurors" were asked to rate the narrative for its quality, it would be rated as a "good" narrative. However, if the narrative were poorly constructed, it presumably would be rated of low quality. Important to note, the primary hypothesis of this chapter is that given that the facts are constant across different narratives, a well-constructed narrative should provide a relatively high rating of guilt, that is, be in agreement with the prosecutor's position, whereas with a poorly constructed narrative, a relatively low rating of guilt should be obtained. In other words, if the prosecuting attorney provides a good narrative, a jury or whoever does the ratings should be likely to agree with the prosecuting attorney with a guilty verdict; whereas if the prosecutor's narrative is poor, the guilty rating will be lower.

We have argued that the guilt ratings should be in agreement with the quality of the narrative of the prosecuting attorney. But how do you tell whether you have a narrative of high or low quality? Fortunately, there has been research on this issue. A number of historians were asked what characteristics they thought were found in a good narrative (Leinhardt, Stainton, Virji, & Odoroff, 1994). The following characteristics were indicated: coherence, completeness or exhaustivity (in the sense of using all available, related information), causality, chronology, and contextualization (the contents of a given narrative relate to a broader context of time and place). In addition, Pennington and Hastie (1993), in their jury decision work, emphasized coherence, and also mentioned coverage (quite similar to completeness), uniqueness (which would suggest the highest quality narrative should be judged better and different from others), and goodness-of-fit.

Given the characteristics of a good narrative as described previously, it was hypothesized that if a narrative lacked one or more of these characteristics, as for example having relatively poor coherence, then the quality rating of that narrative should be relatively low. Further, as mentioned earlier, the rating of the accused's guilt should be lower than when the narrative has good coherence. The present study examined the impact of the quality of a prosecuting attorney's opening or closing narrative statement on juror acceptance of the statement and of the accused's guilt. Moreover, the jurors can be influenced by at least two factors, the facts of the case and the narrative that includes the facts. If, however, the facts are held constant, but the narrative is disrupted by a manipulation of one or more of the criteria for good narrativity, then (a) judgments of narrative quality and persuasiveness should be lower than for a condition with no narrative disruption, and (b) given a, guilty ratings should be less in agreement with the prosecutor's side of the case, because the narrative is written as support for the prosecutor's position.

Four hypothetical murder cases were employed with four respective manipulations of narrative content. One condition, the *baseline* condition, consisted of presentation of the baseline or standard narrative. In a second condition, items were deleted making it an incomplete narrative as compared to the baseline. This manipulation would test the completeness criterion of narrative quality and is termed the *incomplete* condition. In a third condition, while maintaining local coherence, the text was scrambled, although the contents were the same as the baseline version. This manipulation was designed to reduce coherence and chronology, two criteria of good narrativity. Both of these factors, coherence and chronology, were manipulated together because disrupting one typically disrupts the other. This manipulation constitutes the *coherence/chronology disruption* condition. The fourth condition was the *causation disruption* condition. The changes made in this condition were designed to soften the causal emphasis of the narrative by using wording and content that would produce some doubt concerning the defendant's guilt. The specific manipulations are presented in the next section.

An additional variable, manipulated before the reading of each narrative, consisted of presenting one half of the participants with an evidence list "obtained from witnesses" that provided information about the particular case. The reason for including this manipulation was primarily in relation to the incomplete condition. If the participants were to be influenced by having incomplete information, then they would need to notice that the text was incomplete because some information they had been given would not be in the text.

EXPERIMENT

Method

Participants. Sixty-four college students participated in the basic experiment, with 16 serving in each of four conditions.

Design and Materials. The participants were asked to assume the role of a juror and to read the summary presented by the prosecuting attorney for each of four cases as they would find in a courtroom. Four narrative conditions were employed that were described in the preceding section. Four narratives were constructed, each having four versions. One version of each narrative was assigned to each of the aforementioned four conditions via a 4×4 greco-latin square. Each participant thus served once in each condition and was presented with each narrative one time.

As noted, four texts were used. One text and its changes in the various conditions is presented here to show the manipulations that were performed. The remaining two stories that were employed (one was eliminated) and their changes are found in Appendixes A and B. The baseline condition text for "The Car Accident" was as follows:

> The victim, Roger Wilson, had dropped off his coworker, Susan Walker, at her home. He then was driving on Crawford Street in order to get to the freeway. As he was driving, a 6-year-old girl, Marjorie Moran, ran out from behind a parked car. Before Roger could stop, his right fender hit her and she fell to the ground. He quickly got out of his car to check on her and found that she was not seriously injured. Despite this fact, a number of neighborhood teenagers, who were standing nearby, began to push him around, saying things such as "Don't you know how to drive?" Then someone from the crowd took a baseball bat and hit Roger in the head, killing him. This action was seen by a resident living across the street from the altercation, but he was unable to identify who had used the bat. When the police got to the scene of the crime, they took statements from several witnesses, and looked for the bat. In a few minutes, they found a baseball bat in the back seat of a car that was parked

nearby. The car belonged to Matthew Moran, the girl's older brother. Matthew Moran had been among the crowd that attacked Roger Wilson. He was very protective of his younger sister, and sometimes got into fights with people he determined were trying to hurt her. Analyses later revealed that the victim's blood and hair were on the baseball bat. This evidence indicated that Matthew Moran's bat must have been the bat used to hit Roger Wilson. Furthermore, Matthew's were the only fingerprints found on the bat. Matthew Moran claimed that his fingerprints were on the bat because he had used it earlier in the day to play baseball, but playing baseball could not have placed the victim's blood and hair on the bat. The evidence indicates that Matthew Moran's bat must have been used to hit Roger Wilson, and because there were no fingerprints on the bat besides those of Matthew Moran, he must have been the person who hit Roger Wilson with that bat. Matthew Moran, who had the motive, the means, and the opportunity, is guilty of klling Roger Wilson.

In the causality disruption condition, the changes made are given next. The first part of each item is what was in the original baseline text and the second part states the modified contents in the causality text. The basic facts of evidence were maintained. However, the phrasing is designed to reduce the causal impact of the text. Three changes softening the account were numbers 4, 6, and 7. Change number 5 provided an alternative interpretation to the evidence against the defendant. Changes 2 and 3 made some case-related information implicit, whereas Change 1 was a criticism on the victim's character:

1. He quickly got out of his car to check on her and found that she was not seriously injured.→ He quickly stepped out to check on his car and found that it was not damaged.
2. Despite this fact, a number of neighborhood. ...→ A number of neighborhood. ...
3. This evidence indicated that Matthew Moran's bat. ...→ Matthew Moran's bat. ...
4. Matthew Moran's bat must have been the bat used to hit Roger Wilson. ...→ Matthew Moran's bat must have come into contact in some way with Roger Wilson. ...
5. Matthew's were the only fingerprints found on the bat. ...→ Matthew's were the only fingerprints found on the bat, indicating that he had touched it and, that no one else could have touched it, unless they were wearing gloves. ...
6. ... had used it earlier in the day to play baseball, but playing baseball could not have placed the victim's blood and hair on the bat→ ... had used it earlier in the day to play baseball.

7. bat must have been used ... must have been the person. ...→ bat was probably used ... he was probably the person. ...

The omissions made as follows were those parts of the text deleted in the incomplete condition. Deletions 5, 6, and 7 could have had a bearing on the evaluation of the evidence:

1. his coworker
2. He quickly got out of his car to check on her and
3. but he was unable to identify who had used the bat.
4. When the police got to the scene of the crime, they took statements from several witnesses, and looked for the bat.
5. In a few minutes, they found a baseball bat in the back seat of a car that was parked nearby.
6. The car belonged to Matthew Moran, the girl's older brother.
7. Matthew Moran claimed that his fingerprints were on the bat because he had used it earlier in the day to play baseball.

In the coherence/chronology disruption condition, the numbers of the baseline sentences that provided the order of sentences were: 7, second half of 17, 4, 3, 1, 5, 11, 13, 15, 14, first half of 17, 9, 10, 12, 16, 8, 6, 18, 19. This order with minor sentence modifications provided local coherence but took away chronological order. The contents, however, were identical to the baseline condition.

In the preliminary witness information condition, the list provided was "Body of victim found in the middle of a suburban street"; "Eyewitness accounts by neighbors: saw teenager attacking victim"; "Eyewitness account by one neighbor: saw an unidentified person hitting the victim with a baseball bat"; "Baseball bat: found in back seat of nearby car, belonged to Matthew Moran, victim's blood and hair present; also Matthew Moran's fingerprints."

Measures. After reading each narrative, each participant was asked to provide a rating on each of five 1–10 scales and answer three open-ended questions. The question of the first scale was "Do you think the accused is guilty?" (1 = definitely guilty, 10 = definitely not guilty). These numbers were reversed in the presentation of the data for ease of comparison with other data. The second question was "How confident are you in your decision?" (1 = not confident, 10 = very confident). The third question was "Please rate the overall quality of the summary statement" (1 = not at all good, 10 = very good). The fourth question was "How convincing or persuasive was this statement?" (1 = not at all convincing, 10 = very convincing). The fifth question was "How good of an argument did the lawyer make for the case?" (1 = not at all good, 10 = very good). Questions 3, 4, and 5 were all concerned with the narrative quality. The emphasis in each was different, however, with the respective foci on overall quality, persuasiveness, and the quality of the lawyer's construction. Subsequently, the three open-ended questions presented were: "What was good about this argument?" "What was missing?" "How could this statement be improved?"

Results

Overall, the mean guilty rating (with 1 = definitely not guilty and 10 = definitely guilty) was equivalent for three of the four texts, 7.5, 7.5, and 7.4, respectively. The results of the fourth text yielded a less guilty mean and a different distribution of ratings than found in the other three texts, with six participants providing <4 not-guilty ratings, whereas in the other three text conditions there were one, one, and zero not-guilty ratings and only three >6 guilty ratings compared to 9, 10, and 14 for the other three texts. This fourth text was therefore deleted from analyses.

The first row of Table 14.1 presents the mean guilty rating for the four disruption conditions. As shown, the baseline and incomplete conditions showed the highest guilty ratings, with the causation disruption and coherence/chronology disruption conditions showing lower ratings. Comparisons via one-tailed t tests indicated the baseline judgments were significantly greater for the baseline than the coherence/chronology, $t(94) = 2.68$, $p < .005$, and the causation disruption conditions, $t(94) = 2.11$, $p <$

TABLE 14.1

Mean Ratings for the Experimental Conditions

	Condition			
	Baseline	*Incomplete*	*Causation Disruption*	*Coher/Chron Disruption*
Guilty Rating	8.0	7.9	7.0	6.6
Confidence	8.0	7.6	7.4	7.8
Narrative				
Overall Quality	7.8	7.8	6.9	4.9
Convincingness	7.6	7.9	6.9	5.3
Argument Stated	7.6	7.9	6.9	5.1

.015. The second row shows the mean confidence held for the guilty ratings. The greatest confidence was in the baseline condition, followed by the incomplete, causation disruption, and the coherence/chronology disruption condition. Confidence was generally high, although the baseline condition yielded significantly greater confidence than the causation disruption condition, $t(94) = 1.65$, $p < .05$.

The next three rows of Table 14.1 refer to judgments of the narrative. Row 3 presents the mean judgment for the overall quality of the narrative. The baseline and incomplete conditions again scored highest with the coherence/chronology disruption condition mean found considerably below the baseline mean, $t(94) = 4.70$, $p < .005$, and the causation disruption mean somewhat below that of the baseline, $t(94) = 2.01$, $p < .025$. For the next data row, convincingness, the disruption coherence/chronology condition mean was considerably below that of the baseline, $t(94) = 5.60$, $p < .005$, with the disruption causation condition mean also below, $t(94) = 1.64$, $p < .05$. The fifth row data indicate that the mean ratings for the strength of the prosecutor's argument produced the same results as the narrative quality and convincingness judgments. The baseline was significantly greater than both the disruption coherence/chronology, $t(94) = 4.15$, $p < .005$, and the disruption causation, $t(94) = 2.21$, $p < .015$, conditions.

The results presented thus far indicate that the disruption in the causation and coherence/chronology conditions produced poorer judgments of narrative quality than found in the baseline condition, thereby supporting the hypothesis that these particular disruptions would produce such ratings. Furthermore, for these two conditions, guilt ratings were also lower than those found in the baseline condition, supporting the hypothesis that poorer narrative quality would be related to lower guilt ratings. The incomplete condition, however, produced neither lower narrative quality ratings nor lower guilt ratings than the baseline condition.

The results for the coherence/chronology disruption lend strong support to the idea that coherence is critical to narrative quality (Klettke, 1997; Leinhardt et al., 1994; Pennington & Hastie, 1993). The results of the causation condition are in some way more interesting, however, because although the factual evidence was maintained in both the coherence/chronology and causation conditions, the manipulation in the disruption causation condition consisted of making a number of issues less certain.

The incomplete condition results were compared to the incomplete condition plus the evidence list condition. None of the five comparisons shown by the rows of Table 14.1 approached significance. The representative ratings for the Row 1 to the Row 5 measures for the text alone and text plus evidence list conditions, respectively, were: 6.4 and 6.3, 7.5 and 8.1, $p < .025$, 6.6 and 7.0, 6.6 and 7.1, and 6.6 and 7.0. These findings suggest that even when initially given a list of evidence items before reading the narrative, omitting text contents made no difference in the judgments, except to provide more confidence in the guilt ratings. The results therefore do not support completeness as a narrative characteristic that influences judgments of narrative quality or of guilt.

The results for the condition having the evidence list plus text were also comparable to the incomplete condition in the combined other three narrative conditions. The means over the five row measures of Table 14.1 for the incomplete condition alone and the incomplete condition plus information conditions, respectively, were: 6.8 and 6.9, 7.3 and 7.9, 5.2 and 5.6, 7.4 and 7.4, 7.6 and 7.5, and 7.6 and 7.7.

Open-Ended Responses. The open-ended questions yielded results that supported the ratings and provided a type of manipulation check. In the coherence/chronology disruption condition, 23 statements were made indicating that the prosecutor's statement lacked coherence, did not make sense, was hard to follow, or jumped around. In contrast, no one made such statements in the baseline disruption causation, or incomplete conditions. In the disruption causation condition five people stated, "You should not use the word *probably* so much." In contrast, this type of comment was never stated in the baseline condition. This finding supports the weakening of causality attempted in the causation disruption occurred in the way intended via the manipulation. In the incomplete condition, however, participants did not comment on any missing information. Instead, participants in all conditions made statements involving the need for witnesses or for proof. In sum, the answers to the open-ended questions were consistent with the rating data.

GENERAL DISCUSSION

The present study supports the idea that a narrative under particular circumstances may be viewed as a component of argument; that is, a narrative can be used as a reason that supports a claim in much the same way that a reason supports a claim in an enthymeme. Further, the present study distinguishes between support for a claim coming from the facts supplied in the narrative and the story created by weaving those facts together. With the facts held constant, the quality of the story was manipulated, and this was related to judgments of narrative quality and guilt ratings.

The coherence/chronology and causation disruptions yielded significantly poorer ratings of narrative quality and lower guilty ratings than the baseline condition. But because the narratives in these three conditions contained the same facts, the results need to be attributed to how the critical information was presented. The rating results in the coherence/chronology disruption condition taken with the open-ended question results suggest participants were quite sensitive to the disruption of chronological order and lack of coherence found in this condition. These findings also support the historians' emphasis on coherence (Leinhardt et al., 1994) and the conclusions about jury decision making stated by Pennington and Hastie (1993). Similarly, Klettke (1997) found that coherence in statements (noncontradictory vs. contradictory) was more important in judgments of guilt and plausibility than was the amount of evidence. Another type of explanation for the coherence/chronology results, suggested by Lorraine Higgens via personal communication (March, 1998), is that ethos may have lost its power. In other words, the prosecuting attorney has an authority role, but if she or he presents a somewhat incoherent narrative, the respect for that attorney's competence would likely be lessened, and the rating of the convincingness and quality of argument would be relatively low. Correspondingly, the juror would likely be unconvinced by the presentation and would provide a relatively low guilt rating.

The causation disruption manipulation also produced significant differences of narrative and guilt ratings when compared to the baseline findings. As stated, the changes made in the causation disruption condition were designed to soften the causal emphasis of the narrative by using wording and content that would produce some doubt concerning Matthew Moran's guilt. Similarly, in the causation condition of the other two narratives, the Hunting Incident and the Window Shooting, the change to a more probabilistic account is at least comparable to that of the Car Accident narrative, as shown in the Appendixes. The causation disruption condition results thus suggest that narrativity ratings and guilt ratings may be influenced by the use of probabilistic wording or wording that reduces causal certainty, especially when such wording provides a sense of doubt about some aspect of the accused's role in committing the crime. Moreover, this happens even though the facts of the case are included in the narrative.

A point of contention could be that the content deletions may have had something to do with the low ratings in the causation disruption condition, especially if the content that was deleted was critical to the evidential statements of the case. However, Table 14.2 outlines the basic elements of the case against the defendant, and all of the information present in Table 14.2 is present in all conditions. Thus, the components of the prosecutor's case were constant. Similarly, although ratings of coherence were not obtained, a reading of the causation disruption text contents suggests that at least on intuitive grounds the disruption causation condition was as coherent as the baseline.

The failure to find significant effects in the incomplete condition raises the question of how much and what would need to be deleted from a narrative for the narrativity and guilt ratings to be affected. Either enough material would likely need to be removed to reduce coherence or critical facts would need to be deleted such that doubt would occur concerning the guilt of the accused. This possibility points to the likelihood that a high-quality narrative could be constructed that would omit some critical facts, and that individuals would rate narrativity high but guilt low. In other words, narrative quality may be a necessary but not sufficient condition for high guilt ratings. Or, taking the point one step further, the present data do not answer the question of whether a prosecuting attorney could take a case with poor factual evidence and produce a high-quality narrative that would yield a guilty verdict, or whether a defense attorney could construct a high-quality narrative for the defendant that could override a substantial body of evidence. (Another question raised by the present findings is whether a person listening to a prosecuting or defense attorney will construct evaluations of the possible emotional status of the accused and how these influence judgment; see Özyürek & Trabasso, 1997.)

The present findings are related to those of Kuhn (1991), who asked people of various age and education levels questions such as why so many prisoners returned to prison after they were released. One type of evidence she obtained was defined as pseudoevidence, in which case a person did not

TABLE 14.2

Cases for and Against Matthew Moran

Prosecution Case	
	Motive - R.W. hitting sister, fights in past, overprotective
	Means - His bat: in his car, his fingerprints, victim's hair
	Opportunity - Opportunity to obtain bat; was in group
Defense Case	
	No one saw him hit the victim (not stated)
	Someone else could have been wearing gloves
	Fingerprints on bat from playing baseball

differentiate a description of the phenomenon in question from its explanation, and within this category, many of the answers were stories or narratives. This result suggests that when individuals are not aware of specific causes or perhaps are not trained to analyze issues in a specific cause–effect manner, they resort to telling narratives to support their point. Indeed, telling narratives to make a point has long historical roots and a wide range of application, as in literature, songs, folklore, and other phenomena.

The present study suggests that how facts are integrated and presented is important, and of special importance is the coherence and causal chain provided in the narrative. In the cognitive literature the narrative is privileged in a number of ways. Narratives are apparently easier to read, write, summarize, and remember than other text forms (Mandler & Johnson, 1977; Stein & Glenn, 1979), and the present study suggests that good narrative form can also influence the plausibility and persuasiveness of the information presented.

REFERENCES

Angell, R. B. (1964). *Reasoning and logic.* New York: Appleton–Century–Crofts.

Collingwood, R. G. (1946). *The idea of history.* London: Oxford University Press.

Cronon, W. (1992). A place for stories: Nature, history, and narrative. *Journal of American History, 78,* 1347–1376.

Epstein, T. L. (1996). Urban adolescents' historical understanding: Differences in Black and White. *Theory and Research in Social Education, 26,* 399–402.

Foucault, M. (1972). *The archeology of knowledge.* (A. Sheridan Smith, Trans.). New York: Pantheon. (Original work published 1969)

Graesser, A. C., Singer, M., & Trabasso, T. (1994). A constructionist theory of inference generation during narrative text comprehension. *Psychological Review, 101,* 371–395.

Kintsch, W. (1998). *Comprehension.* Cambridge, England: Cambridge University Press.

Klettke, B. (1997). Repressed memories and testimony evaluation (Undergraduate research project). Memphis, TN: Memphis University.

Kuhn, D. (1991). *The skills of argument.* Cambridge, England: Cambridge University Press.

Leinhardt, G., Stainton, C., Virji, S. M., & Odoroff, E. (1994). Learning to reason in history: Mindlessness to mindfulness. In M. Carretero & J. F. Voss (Eds.), *Cognitive and instructional processes in history and the social sciences* (pp. 131–158). Hillsdale, NJ: Lawrence Erlbaum Associates.

Mandler, J. M., & Johnson, N. S. (1977). Remembrance of things parsed: Story structure and recall. *Cognitive Psychology, 9,* 111–151.

McGuire, M. (1990). The rhetoric of narrative: A hermeneutic critical theory. In B. K. Britton & A. D. Pellegrini (Eds.), *Narrative thought and narrative language* (pp. 219–236). Hillsdale, NJ: Lawrence Erlbaum Associates.

Mink, L. O. (1987). *Historical understanding.* Ithaca, NY: Cornell University Press.

Özyürek, A., & Trabasso, T. (1997). Education during the understanding of narrative. *Discourse Processes, 23,* 305–335.

Pennington, N., & Hastie, R. (1993). The story model for juror decision making. In R. Hastie (Ed.), *Inside the juror. The psychology of juror decision making* (pp. 192–221). Cambridge, England: Cambridge University Press.

Stein, N. L., & Glenn, C. G. (1979). An analysis of story comprehension in elementary school children. In R. O. Freedle (Ed.), *New directions in discourse processing* (Vol. 2, pp. 53–120). Norwood, NJ: Ablex.

Sturken, M. (1997). Reenactment, fantasy, and the paranoia of history: Oliver Stone's docudramas. *History and Theory, 36,* 64–79.

Thucydides. (1972). *History of the Peloponnesian War.* (R. Warner, Trans.). London: Penguin.

Trabasso, T., van den Broek, P., & Suh, S. (1989). Logical necessity and transitivity of causal relations in stories. *Discourse Processes, 12,* 1–25.

Wertsch, J. V., & Rozin, M. (in press). The Russian Revolution: Official and unofficial accounts. In J. F. Voss & M. Carretero (Eds.), *Learning and understanding in history.* London: Woburn.

White, H. (1987). *The content of the form.* Baltimore: Johns Hopkins University Press.

Wineburg, S. S. (1991). On the reading of historical texts: Notes on the breach between school and academy. *American Educational Research Journal, 28,* 495–519.

APPENDIX A

PROSECUTING ATTORNEY'S SUMMARY—THE HUNTING INCIDENT (BASELINE)

Paul Ryan and Rob Brown drove to DuBois together in Rob's truck. For years they had driven there at the beginning of deer season and hunted together. During this hunt, however, Rob Brown was shot and killed by Paul Ryan. According to Paul, as he and Rob walked approximately 20 yards apart, he saw the head and antlers of a deer to his right. He then raised his gun, aiming at the deer, and fired. He claims that his shot missed the deer, but that he saw something else fall. He then ran to where this had taken place to see what he had hit, and found Rob lying in the grass, dead. Paul says that, fraught with grief, he immediately went to a nearby cabin owned by Jack Haskins where he contacted the police. He claims that this was all a terrible accident. However, the police investigation concluded that the area where Paul and Rob had been walking had only a few trees and a little brush and thicket, and that most likely, Rob was fully visible to Paul. Furthermore, during the investigation, Rachel Ryan (Paul's wife) testified that she and Rob had been having an affair for over a year and a half. This was corroborated when Mary Jacoby, a friend of Rachel Ryan's, testified that Rachel had told her about the affair over a year ago. Mary also testified that directly prior to the fatal hunting trip, her conscience got the better of her and she had gone to Paul and told him all about his wife's affair with Rob. Considering the fact that the police determined that Paul must have been able to see Rob with no difficulty when he was shot, and the fact that Paul had just discovered that Rob had been having an affair with his wife, it is clear that the death of Rob Brown was a premeditated murder, and not a freak accident. The discovery of the illicit affair was more than enough to enrage Paul, motivating him to kill. Paul should be found guilty of first degree, premeditated murder in cold blood.

Causal Version

- He claims that his shot missed the deer ...→ His shot missed the deer ...
- Paul says that, fraught with grief, he immediately went to a nearby cabin ...→ Paul went to a nearby cabin ...
- However, the police investigation concluded that the area where Paul and Rob had been walking had only a few trees and a little brush and thicket, and that most likely, Rob was fully visible to Paul.→ *not included in causal condition*
- Paul must have been able to see Rob with no difficulty ...→ Paul might have been able to see Rob with no difficulty ...
- Paul had just discovered that Rob had been having an affair with his wife ...→ Paul had probably just found out from Mary that Rob had been having an affair with his wife ...

Incomplete Version

- in Rob's truck
- owned by Jack Haskins
- This was corroborated when mary Jacoby, a friend of Rachel Ryan's, testified that Rachel told her about the affair over a year ago.
- Mary also testified that directly prior to the fatal hunting trip, her conscience got the better of her and she had gone to Paul and told him all about his wife's affair with Rob.

Sentence Number in the Baseline Condition That Provided the Order in the Incoherent Version. (numbered 1-16 in baseline) 4, 2, 9, 11, 1, 8, 10 (first half), 12, 3, 10 (second half), 7, 5, 14, 13, 6, 15, 16.

APPENDIX B

PROSECUTING ATTORNEY'S SUMMARY—THE WINDOW SHOOTING (BASELINE)

Sam Fogarty, a man of 62 years, was sitting in the study, reading. It was about 11:00 P.M. and as he got up, probably to leave the room, he heard a noise at the window. He walked to the window, opened it, and saw a revolver pointed at him. The gun was fired, and Sam fell backwards into a crumpled heap. Sam's wife, in another part of the house, heard the shot. She came quickly into the study and saw Sam's body. She screamed and then went to the body, feeling for a pulse. There was none. She then phoned the police and they arrived within 10 minutes. The homicide team, led by Detective Ruskin, was notified and arrived 15 minutes later. A thorough analysis of the murder scene was conducted.

A gun was found outside the window and a ballistics analysis indicated that it was the gun used to commit the murder. Furthermore, fingerprints were found on the gun that belonged to Avery Haskins, the victim's next-door neighbor. It was also established that the gun in question was registered to Avery Haskins, and was seen in his possession as recently as 2 days before the killing. Detective Ruskin interrogated other neighbors, and they indicated that there had been an ongoing feud between Fogarty and Haskins, with Haskins claiming that Fogarty had gotten Haskins fired from his job at the firm where they were both accountants. Fogarty claimed that there were no bad feelings between the two, and that they were, in fact, good friends.

Detective Ruskin also found that a neighbor living across the street, Minerva Polson, had her dog out for a walk the night of the killing at about 10:55 P.M., and she noticed a person that she identified as Avery Haskins was running from the Fogarty house. Avery Haskins was the owner of the murder weapon, he had motive and opportunity to kill, and he was seen heading from the crime scene just prior to the murder. Avery Haskins is guilty of premeditated murder.

Causal Version

- Fogarty claimed that there were no bad feelings between the two, and that they were, in fact, good friends.→ *not included in causal version*
- ... Avery Haskins was running from the Fogarty house.→ Avery Haskins coming out of his house, turning in the direction of the Fogarty house.

- … he had motive and opportunity to kill …→ … he may have had motive and opportunity to kill …
- … he was seen heading from the crime scene just prior to the murder.→ … he was seen heading toward the crime scene just prior to the murder.

Incomplete Version

- a man of 62 years
- She came quickly into the study and saw Sam's body.
- She screamed and then went to the body, feeling for a pulse.
- There was none.
- … and was seen in his possession as recently as 2 days before the killing.
- Minerva Polson

Sentence Number in the Baseline Condition That Provided the Order in the Incoherent Version. (numbered 1-19 in baseline) 12, 1, 17, 15, 2 (excluding time), 10, 5, 14, 7, 3, 16, 4, 6, 8, 13, 9, 11, 18, 19.

15

The Dandelion War

Hildy Ross, Michael Ross,
Anne Wilson, and Melissa Smith
University of Waterloo

Since 1993, Tom Trabasso, Nancy Stein, and the two senior authors have met several times a year to eat, drink, and talk psychology, especially the psychology of family conflict. We have rendezvoused in Chicago, Waterloo, Colorado, Quebec, and Israel to formulate research and grant proposals. Our joint work examines how young children and their parents understand and resolve conflicts. Conflict is inevitable in even the happiest of families and it is often not settled to the mutual satisfaction of the protagonists (Vuchinich, 1987). We seek to understand why family disputes are so difficult to resolve and to develop procedures that facilitate their successful resolution.

The research described here has benefited enormously from discussions with Tom and Nancy. They have encouraged us to analyze young children's narratives of their conflicts. Children's individual narratives indicate their understanding of the conflicts and may offer clues as to why disputes are difficult to settle. In the present chapter, we describe a study in which we asked young siblings from 40 families to provide detailed narratives of their conflicts with each other. In each sibling dyad, one child was between the ages of 3½ and 5, and the other between 5½ and 8. Our interviews reflected the Stein and Trabasso emphasis on children's goals and emotions in relation to their actions. This enabled us to analyze their conflicts in terms of the justifications and goals that they attributed to themselves and their siblings. We found that the two young protagonists often provided quite different depictions of the same conflict. For example, they reported more harmful actions by siblings than by themselves and offered more elaborate justifications for their own conflict actions (Smith, Wilson, H. Ross, & M.

Ross, 1997; see also Albro, 1996; Baumeister, Stillwell, & Wotman, 1990; Shantz, 1993; Stein & Albro, 1997). In this chapter, we describe our procedures and present an intensive analysis of one sibling dispute to demonstrate how we arrive at such conclusions and to illustrate the strengths and weaknesses of our approach.

Children were interviewed on three separate occasions about a variety of sibling conflicts that were reported by their parents. Parents were asked to call our phone diary line daily during the week prior to the children's interviews and to report on any conflicts that had occurred between the children. We asked for brief accounts of what the conflict was about, who was involved, how it started, progressed, and ended, and how important the conflict was for the children. We then selected the most significant conflict for the children, according to the parents' reports, and interviewed the children individually, in every case within a week of the event reported by the parents. We asked the children how the problem started and continued, what each party wanted and why, how each felt and why, how the conflict ended, and whether the outcome was fair. Based on the parents' descriptions in the phone diaries, the interviewers set the scene for the children's narratives. The key was to provide enough detail so that both siblings would recognize and retell the same event while providing as little direct information about the conflict as possible. In our example dyad, Katie is 4 years 2 months and Jesse is her 5½-year-old sister. In describing the Dandelion War, Katie and Jesse refer to another sibling, their younger brother Daniel. Here is how the interviewers introduced the sample sibling conflict to each child:

Interview With the Younger Sibling

Interviewer:	This week, I want you to tell me about, um, sometime this week when you and …
Katie:	Jesse?
Interviewer:	And Jesse and Daniel and Mom and Dad were over, and Mom and Dad were playing tennis. And then you guys were going to go to the little park next to the tennis place. And Jesse picked up a dandelion flower. Do you remember that?
Katie:	Yeah.
Interviewer:	Yeah? What happened?
Katie:	We both got, Daniel wanted to pick flowers for Mommy and Daddy.
Interviewer:	Oh, Daniel wanted to pick flowers for Mommy and Daddy? And then Jesse picked up a flower and what happened?
Katie:	And then Jesse started um, started the fighting.

Interview with the Older Sibling

Interviewer: The first one that I'm going to ask you about is something that happened between you and your sister Katie. And I want you to tell me about this week when your mom and dad were playing tennis. Do you guys go to a tennis place where they play?

Jesse: (nods head yes)

Interviewer: And you and Katie went to the little park that was near there.

Jesse: What, the little, the little one by the baseball game?

Interviewer: Yeah. I think. Oh, is there one by the tennis? Where your mom and dad play tennis? Is there a field where you go play in the grass?

Jesse: (nods head yes)

Interviewer: Okay.

Jesse: There's one by the baseball park and there's one way out by, over there.

Interviewer: Oh, okay. Well this is when you and Katie were going to one of the parks. And you were picking dandelions in the field. Do you remember that? Do you remember what happened? When you were picking dandelions?

Jesse: I don't know if I was picking dandelions.

Interviewer: You don't remember if you were picking dandelions? Did you uh pick a dandelion and then, um, did you drop it?

Jesse: Yeah.

Interviewer: And then what happened?

Jesse: We started to fight.

The idiosyncrasies in the interviewers' introductions to the conflict probably reflect their own personal styles as well as those of the children. In this dyad, the younger child, Katie, is more effusive than her older sibling throughout the interviews; Katie needs fewer prompts to get started on her narrative. The interviewers begin by describing the physical setting and introducing the object of contention—the dandelion. The younger sibling picks up the story from there—"Jesse started … the fighting." The older child first expressed some doubt but eventually launched into the story. But Jesse's story begins differently from Katie's. Whereas Katie blamed Jesse for starting the dispute, Jesse claimed that they both started the argument. Our first goal in the analysis of this conflict was to determine the course of the actions each child reported. Descriptions of actions come in response to our initial questions about what happened and what happened next, but action descriptions are also embedded in children's responses to later questions about goals, emotions, and fairness.

We are interested in contrasting the children's narrations of the same conflict. To facilitate a point-by-point comparison, we extracted action descriptions from the full text of the children's transcripts and ordered them in an approximate chronological sequence, according to the versions of the event as described by both children and the parent.

The chronological narratives can be divided into four major scenes or episodes: (a) Dandelion Changes Hands, (b) Physical Fight, (c) Tattling and Intervention, (d) Reconciliation. Although details differ, both children's stories contain these scenes and the two children are clearly describing the same conflict (Table 15.1). They agree that the conflict began over a dandelion that Jesse had picked and dropped, and that Katie had gained possession of it. They agree that some fighting did occur, that one child rolled the other on the ground, that someone cried, that they went to their father who talked to them, and that they reconciled afterward by hugging and kissing. Taken individually they both seem to remember the conflict well, and to agree on the major elements of the story. But their tales also differ in important ways. Katie makes clear that she had the flower in her possession and claims that Jesse grabbed it from her. Jesse states that she was starting to retrieve the flower herself when Katie grabbed it out of her hand. She denies saying anything to her sister about grabbing the flower; whereas, Katie explicitly describes Jesse's excluding her from being able to have some of the flowers, denying her any more flowers, and threatening physical aggression. In Katie's version, Jesse got the flower back and "just throwed it"; and, in Jesse's version, Katie would not return the flower. According to Katie, Jesse started the fighting: Jesse rolled her on the grass, and she, Katie, did nothing in response. According to Jesse they both started fighting, Katie rolled her on the grass, and hit, tripped, and kicked her. Each claims that she cried, and Jesse explicitly denies that Katie had done so. Katie has a far more vivid report of what her parents did or said when they intervened, and of the punishment Jesse was to receive. Jesse, for her part, denies getting into any trouble. There are differences in the reports of the reconciliation as well: Katie reports that Jesse apologized and kissed her; Jesse emphasizes that they both apologized.

These versions are not just different from one another. The children represent their own perspectives on the same event, and each recounts a conflict in which she was less blameworthy, more justified in her actions, and suffered more severely than the other. The younger sister claims parental support for her version of the story. Table 15.2 begins the process of unraveling the perspectives by coding the transgressions that each child attributed to self and other in her narrative. We apply parallel categories to the transgressions that the children describe as we have previously used to code observations of sibling conflict (den Bak & H. Ross, 1996; H. Ross & den Bak-Lammers, 1998; H. Ross, Filyer, Lollis, Perlman & Martin, 1994). We attempt to code each action taken against the sibling as a separate

TABLE 15.1

Action Transcript of the Dandelion War
According to the Accounts of the Two Children

Younger Sibling's Account	*Scenes*	*Older Sibling's Account*
I just took one. She just, she just dropped one. And then I got it. I started picking up the flower.	**Dandelion Changes Hands**	[Did you uh pick a dandelion and then, um, did you drop it?] Yeah. And I dropped it.
[Oh. Did you have the flower?] nods head yes. Jesse, when I had it, Jesse just grabbed it out of my hands. And then when, when she did that she just grabbed it …		I was starting to grab it first … … but then Katie just grabbed it out of my hand … Cuz Katie grabbed the flower. … cuz she wouldn't give it back.
… and then just, she just throwed it.		[What happened to the dandelion?] Got lost.
She said "These are only for Mommy, Daddy, and Danny." And then Jesse say, "You're going to get no more flowers or I might roll you over on the grass."		[What did you say? Did you say anything to her when she went to grab the flower?] Shakes head no. [Did you say: "Hey that's mine" or anything like that?] Shakes head no. [No? You never said anything? Did you say "Don't" or "Don't pick that, it's mine" or?] Shakes head no.
And then Jesse started um, started thc fighting. [Started what?] Fighting. But, um, I didn't started it. Jesse [started it].	**Physical Fighting**	We started to fight. … and then we started to fight.
Jesse, but Jesse was, was rolling me on the, Jesse was trying, um, to roll me on, on, on, on the grass. She was trying to roll me over. Cuz, cuz she started rolling me over.		Cuz Katie was rolling on me and then, then …
[So she started it. I see. And did you do anything to Jesse after she started it?] Shakes head no. [No? Did you hit her?] Shakes head no. [No? So you didn't do anything at all?] Shakes head no.		[both were] Hitting, and, tripping. Cuz, cuz um, when we were hitting and kicking … … she hit me back and she finished it. Cuz she hit me back. then she kicked me in the head.
I cried.		Katie wasn't crying, I was. Cuz I was crying.

Table 15.1 continued

Table 15.1 continued

Younger Sibling's Account	*Scenes*	*Older Sibling's Account*
She said um, she, I told on Jesse.	**Tattling & Intervention**	I had to go to Daddy. We both did.
Um, Daddy said "Come here." And he said "What happened?"		Daddy talked. [What did he say?] I forget.
And I said "Jesse, um, Jesse started it first!"		[Think hard. Can you remember anything that he said?] No. I can't remember.
And Jesse said "Katie started it first!" And I said "No, Jesse did!"		
And she, and dad, and mommy saw um, me, Jesse sawed, mommy, my mommy sawed Jesse rolling me over. So she started it.		[Do you remember what she (Mom) said or did?] Nothing.
When they went, we were going home and Jesse went to her bed.		[When he talked to you. Did you guys get in trouble?]
		No.
[Mom and Dad said] When we get home, Jesse's, Jesse's going to bed.		[Did you have to go to your room or anything?] No. We weren't at our house.
[Mom and Dad said] Um, um, if you do one more bad thing, you will go to your room.		
She said sorry.	**Reconciliation.**	[Did dad say anything to you guys at the end? That you had to make up? Or that you had to hug or anything?] Uh, yeah.
We gave a hug to each other.		[And what did you have to say to each other?]
And she gave me big kiss.		Sorry and I love you.
		Cuz we both said it. Cuz she said the same thing as me.

Note. Any interviewer questions that were necessary to understanding the children are included in brackets within each child's account.

transgression, but when the children mention what could be the same hit or kick several times, we count it as a single transgression. The children's narratives reveal an extensive range of transgressions. These include violations of (a) property entitlement (taking or attempting to take something possessed or owned by the other), (b) property damage (harming an object claimed by the other, by breaking, throwing, or scribbling on it), (c) bossing (ordering or otherwise attempting to verbally control another's actions), (d) physical harm (physical aggression or control), (e) verbal harm (insulting, name-calling, or threatening another), (f) tattling (informing parents of the other's wrongdoing), and (g) lying (attempting to deceive someone).

TABLE 15.2

Transgressions of Self and Other According to the Accounts of the Two Children

Transgressions According to Katie	*Younger Sibling's Account*	*Scenes*	*Older Sibling's Account*	*Transgressions According to Jesse*
self property entitlement	I just took one. She just, she just dropped one. And then I got it. I started picking up the flower.	**Dandelion Changes Hands**	I was starting to grab it first.	**self property entitlement**
other property entitlement	Jesse, when I had it, Jesse just grabbed it out of my hands. And then when, when she did that she just grabbed it …		… but then Katie just grabbed it out of my hand … Cuz Katie grabbed the flower.	**other property entitlement**
other property damage	… and then just, she just throwed it.		… cuz she wouldn't give it back	**other property entitlement**
other bossing	She said "These are only for Mommy, Daddy, and Danny."			
other bossing	And then Jesse say, "You're going to get no more flowers …			
other verbal harm	… or I might roll you over on the grass."			
other fighting	And then Jesse started um, started the fighting. [Started what?] Fighting.	**Physical Fighting**	We started to fight. … and then we started to fight.	**both fighting**

Table 15.2 continued

Table 15.2 continued

Transgressions According to Katie	*Younger Sibling's Account*	*Scenes*	*Older Sibling's Account*	*Transgressions According to Jesse*
other physical harm	Cuz, cuz she started rolling me over.		Cuz Katie was rolling on me and then, …	**other physical harm**
	Jesse, but Jesse was, was rolling me on the, Jesse was trying, um, to roll me on, on, on, on the grass. She was trying to roll me over.		[both were] Hitting, and, tripping. Cuz, cuz um, when we were hitting and kicking …	**both physical harm**
			… she hit me back and she finished it. Cuz she hit me back.	**other physical harm**
			then she kicked me in the head.	**other physical harm**
self tattling	She said um, she, I told on Jesse. And I said "Jesse, um, Jesse started it first!"	**Tattling & Intervention**		
other tattling	And Jesse said "Katie started it first!"			
other lying	And Jesse said "Katie started it first!" [based on the confirmation of Katie's claim that she ascribes to her mother. "My mommy sawed Jesse rolling me over."]			
self tattling	And I said "No, Jesse did!"			

Our coding of the Dandelion War narratives on these dimensions is shown in Table 15.2. In total, Katie's version includes three of her own and nine of her sister's transgressions. Jesse's account contains three of her own transgressions and seven of Katie's. Our data for the 40 families indicate that older siblings report significantly more transgressions ($M = 6.5$) per conflict than their younger siblings do ($M = 5.3$). Both children cite more transgressions for other (3.6 and 2.8, respectively) than for self (3.0 and 2.5), a difference that was reflected in the Dandelion War narrative as well.

In addition to attributing more misbehaviors to their opponents, children also portray their own transgressions differently from those of their antagonists. Children tend to lessen their own culpability by justifying their actions, arguing that the damage they inflicted was minimal or deserved. Voluntary restitution or apologies could also mitigate blame by making up in some way for the harm done. Conversely, children provide descriptions that maximize the harm done by their opponents and emphasize the malevolence of their intentions. These "maximizing" appraisals have the effect of increasing the culpability of the actor. Overall, our data show that children were significantly more likely to justify their own transgressions than those of their siblings (.96 vs. .39 justifications per transgression) and to maximize the transgressions of their siblings (.25 vs. .10 maximizing statements per transgression).

Jesse and Katie follow the group trends of justifying their own transgressions and maximizing their sibling's. Table 15.3 provides a summary of the justifications and maximizing statements associated with transgressions that each child ascribes to the other in each conflict scene. Some are familiar, being drawn from the children's action descriptions that appear in Tables 15.1 and 15.2. Others come from statements of motivation or emotion that were either spontaneous in the children's narratives or given in response to specific questions about why the children felt, acted, or desired what the narrator ascribed to them. With respect to the first scene, Katie notes that she took *just one* dandelion, implying that there were others available for Jesse, that her ultimate motive was altruistic, that she had prior possession of the flower ("When I had it, Jesse just grabbed it out of my hands"), and that Jesse started the fight. According to Katie, Jesse, too, was altruistically motivated, though perhaps unreasonably so: "She wanted to give every single flower to Daniel." Jesse, on the other hand, provides more justifications for her own actions, insisting that she had the dandelion *first* ("Katie grabbed it out of my hand") and that it was *her* flower. In accidentally dropping the flower, she was not abandoning it or relinquishing it permanently. Furthermore, her motive was altruistic ("cuz I was going to give it

TABLE 15.3

Blaming and Absolving Self and Other

Transgressions According to Katie	*Katie's Blaming and Absolving*		*Jesse's Blaming and Absolving*	*Transgressions According to Jesse*
		Dandelion Changes Hands		
self property entitlement	1. Minimized damage *[I just took one. She just dropped one.]* 2. Altruistic motive *[I wanted to give it to Mommy.]* 3. Possession *[Jesse, when I had it Jesse just grabbed it out of my hands.]* 4. Provocation *[I didn't start it. Jesse um, Jesse started it first.]*	Justifications for Katie's transgressions	none	other property entitlement
	none	Maximizing for Katie's transgressions	1. Negative consequences *[Cuz I was crying.]* 2. No need *[Cuz she was starting to pick them too.]* 3. Unreasonable motive *[Cuz she wanted to grab it.]*	
other property entitlement, property damage, bossing, and verbal harm	1. Altruistic motive *[She wanted to give every single flower to Daniel.]*	Justifications for Jesse's transgressions	1. Possession *[I was starting to grab it first. Katie grabbed it out of my hand.]* 2. Ownership *[My dandelion back. Cuz it was my flower.]* 3. Uncontrolled external cause *[... and I dropped it.]* 4. Altruistic motive *[Cuz I was going to give it to Daniel.]*	self property entitlement
	1. Negative consequences *[I cried.]* 2. Negative consequences *[because I didn't like it.]* 3. Negative appraisal *[cuz she wasn't being very nice.]*	Maximizing for Jesse's transgressions	none	

Table 15.3 continued

Table 15.3 continued

Transgressions According to Katie	Katie's Blaming and Absolving		Jesse's Blaming and Absolving	Transgressions According to Jesse
	4. Negative appraisal *[She was being nasty to me cuz she said "These are only for Mommy, Daddy, and Danny."]* 5. Counters altruistic motive *[She didn't want me to give the flower to Mommy.]* 6. Unreasonable motive *[She wanted to give every single flower to Danny.]* 7. Lack of need *[and then she just throwed it.]*			
		Physical Fighting		
no transgressions described		Justifications for Katie's transgressions	1. Reciprocity *[both started to fight]* 2. Reciprocity *[both were hitting and tripping]* 3. Reciprocity *[She hit me back.]*	other physical harm and fighting
		Maximizing for Katie's transgressions	1. Negative consequences *[I was (crying).]* 2. Negative consequences [(When Katie kicked you in the head, did it hurt?) *Mmhmm* (yes)]	
other physical harm	none	Justifications for Jesse's transgressions	1. Reciprocity *[both started to fight]* 2. Reciprocity *[both were hitting and tripping]* 3. Minimized consequences for other *[Katie wasn't crying.]* 4. Provocation *[Cuz Katie was rolling on me.]*	self physical harm

Table 15.3 continued

Table 15.3 continued

Transgressions According to Katie	*Katie's Blaming and Absolving*		*Jesse's Blaming and Absolving*	*Transgressions According to Jesse*
	1. Negative consequences *[I cried]* 2. Negative appraisal *[Because it wasn't very nice.]* 3. Unprovoked *[I didn't start it. Jesse um, Jesse started it first.]* 4. Sadistic response *[She liked it.]*	Maximizing for Jesse's transgressions	none	
		Tattling		
self tattling	1. Provocation *[because she was being nasty.]* 2. Provocation *[And Jesse said "Katie started it first." And I said "No, Jesse started it."]*	Justifications for Katie's transgressions		no transgressions described
	none	Maximizing for Katie's transgressions		
other tattling, lying	none	Justifications for Jesse's transgressions		no transgressions described
	1. Tattling untrue *[But, um, I didn't start it. And she, and Dad, and Mommy saw um, me, Jesse sawed, Mommy, my mommy sawed Jesse rolling me over. So she started it.]*	Maximizing for Jesse's transgressions		

to Daniel"). Indeed, the justifications offered by each child for this first event of the conflict were remarkably similar. In describing the entire conflict, Katie provides six justifications for her own three transgressions and one justification for Jesse's nine. In turn, Jesse justifies the three transgressions that she attributes to herself eight times, but notes only three justifications for the seven transgressions she attributes to her sister.

Of particular interest, 4-year-old Katie elaborates on a system of hierarchical goals. These goals provide a compelling justification for her own conflict actions, while emphasizing the unreasonable motivation she ascribes to her sister. The conflicting goals and goal hierarchies create cohesion among the diverse actions in Katie's conflict narrative (Trabasso & Stein, 1994; Trabasso, Stein, & Johnson, 1981). First, Katie noted that both she and her sister wanted the dandelion. Katie wanted it to give it to her mother. She didn't break it so that she could give it to her mother. Katie wanted to give it to her mother because Jesse wasn't being very nice. Jesse wanted to take the dandelion from Katie because she didn't want Katie to give it to her mother; Jesse wanted to give every single flower to Daniel. Katie also linked Jesse's physical aggression to the set of goals Katie attributed to her and to the pleasure Jesse took in harming her ("She liked it"). When directly questioned about why Jesse wanted to give every single flower to Daniel, Katie replied that she didn't know. Jesse's own description of her motivation also formed a compelling hierarchy. She wanted her dandelion back because it was her flower and because she wanted to give it to Daniel. In describing Katie's motives, Jesse explains that Katie "wanted a flower because she wanted to grab it." Jesse, in turn, claimed not to know why Katie wanted the flower. Each girl thus casts her own goals as reasonable and justified, and her sister's as arbitrary and incomprehensible, an effect noted when adults describe past conflicts as well (Baumeister et al., 1990).

The children's tendency to stress their own innocence extends to emphasizing the harmful consequences of the other's actions. They both claim that the actions of the other caused them to cry while failing to note that the other also cried. Indeed, Jesse comes close to completely absolving herself of any transgressions by explicitly denying that serious harm had been done—"Katie wasn't crying, I was." Furthermore, they each show limited understanding of the apparently unreasonable motives they attribute to one another, contributing to the impression that both children's acts were unjustified as well as harmful in the other's eyes. Katie notes that she only took one, presumably of the many dandelions that Jesse had, and that Jesse later threw the dandelion away; Jesse states that Katie was picking dandelions herself. These statements indicate that neither child felt the other had any need for this one disputed weed. Katie also makes explicit negative appraisals of Jesse's actions ("cuz she wasn't being very nice" and "because she was being nasty"), which make the negative quality of Jesse's actions clear. Neither of these girls mentioned elements that would maximize her own culpability for transgressions.

To obtain another perspective on the conflicts, we asked a parent to evaluate the children's accounts. During the week between test sessions, we typed each child's report of the conflict. We provided the typed transcripts to the accompanying parent and asked him or her to write comments on the transcripts, clarifying the children's statements where necessary and possible, and evaluating the accuracy of the statements. Recall, as well, that we had obtained an abbreviated version of the conflict from one of the parents in the phone diary. Table 15.4 shows the phone diary version of this conflict (provided by the father) as well as the mother's comments on the typed transcripts of the children's interviews. Thus the phone diary and the commentary were provided by different parents. As it happens in this instance, both parents were near the site of the original conflict. From the phone diary, however, it is unclear what parts of the conflict either of them directly witnessed. At the time, they were playing tennis nearby.

Can we use the parents' accounts and comments to reconcile the major discrepancies in the children's versions? In many cases, one sister mentions some element of the story, for example, the tattling, that is not included in the other's account. In other cases, they provide accounts that directly contradict one another, or the events described by one are categorically denied by the other. Eight clear disagreements are identified in which an explicit discrepancy exists between the children's accounts:

1. Did Katie retrieve the flower from the ground where Jesse had dropped it, as she claims, or did she grab it from Jesse's hand, as Jesse describes? The father sides with Katie; Jesse dropped the flower, Katie picked it up, and then Jesse wanted it back.
2. Did Jesse say any of the things that Katie claims she did? Here both parents are silent.
3. Did Jesse start the fighting, or did they both start? Jesse initiated it according to the father's account.
4. Did Jesse roll Katie on the ground, or did Katie roll Jesse? Dad describes Jesse as tackling Katie, and Mom notes that Katie's account is accurate and Jesse's inaccurate.
5. Did Katie cry, as she claims and Jesse explicitly denies? Both parents agree that she did cry, and Dad, in particular, describes her "screaming and wailing." Does Jesse just miss this?

TABLE 15.4

Integration of the Children's Accounts with the Parents' Description and Review of the Conflict Transcripts

Younger Sibling's Account	*Parents' Commentary*	*Older Sibling's Account*
I just took one. She just, she just dropped one.	**Dandelion Changes Hands**	[Did you uh pick a dandelion and then, um, did you drop it?] Yeah.
And then I got it.		
I started picking up the flower.	*Anyway, Jesse picked up a flower and she dropped it, and then Katie picked it up.*	And I dropped it.
[Oh. Did you have the flower?] nods head yes.		I was starting to grab it first .... but then Katie just grabbed it out of my hand ...
Jesse, when I had it, Jesse just grabbed it out of my hands	*And Jesse wanted the dandelion back again and Katie wouldn't give it to Jesse.*	Cuz Katie grabbed the flower ... cuz she wouldn't give it back
And then when, when she did that she just grabbed it ...		
... and then just, she just throwed it.		[What happened to the dandelion?] Got lost.
She said "These are only for Mommy, Daddy, and Danny."		[What did you say? Did you say anything to her when she went to grab the flower?] Shakes head no. [Did you say: "Hey that's mine" or anything like that?] Shakes head no. [No? You never said anything? Did you say "Don't" or "Don't pick that, it's mine" or?] Shakes head no.
And then Jesse say "You're going to get no more flowers or I might roll you over on the grass."		
And then Jesse started um, started the fighting. [Started what?] Fighting. But, um, I didn't started it. Jessee [started it].	**Physical Fighting**	We started to fight. ...and then we started to fight.

Table 15.4 continued

Table 15.4 continued

Younger Sibling's Account	*Parents' Commentary*	*Older Sibling's Account*
Jesse, but Jesse was, was rolling me on the, Jesse was trying, um, to roll me on, on, on, on the grass. She was trying to roll me over. Cuz, Cuz she started rolling me over.	[Katie Accurate.] [Jesse Inaccurate] *So Jesse tackled her...*	Cuz Katie was rolling on me and then, then,
[So she started it. I see. And did you do anything to Jesse after she started it?] Shakes head no. [No? did you hit her?] Shakes head no. [No? So you didn't do anything at all?] Shakes head no.	*...and [Katie] ended up hitting Jesse.* *...and she ended up hitting Jesse, accidentally probably, but anyway...* [Katie did do something, perhaps pinch.]	[both were] Hitting, and, tripping. Cuz, cuz um, when we were hitting and kicking... ...she hit me back and she finished it. Cuz she hit me back. then she kicked me in the head.
I cried.	[Both children where crying.] *...so of course Katie started screaming and wailing a bit.* *...so they both ended up kind of crying.*	Katie wasn't crying, I was. Cuz I was crying.
	Tattling & Intervention	
She said um, she, I told on Jesse Um, daddy said "Come here." And he said "What happened?" And I said "Jesse, um, Jesse started it first!"	[Katie Accurate] [Jesse Accurate] *... so they came back to the tennis courts and came to see me and I told, I asked Katie—no first I asked Jesse what happened so Jesse told me the story and then I asked Katie what happened and Katie told me her story.*	I had to go to daddy. We both did. Daddy talked. [What did he say?] I forget.
And Jesse said "Katie started it first!" And I said "No, Jesse did!" And she, and Dad, and Mommy saw um, me, Jesse sawed, Mommy, my mommy sawed Jesse rolling me over. So she started it.	*And it was difficult for them to not talk while the other one's talking. I tried to make a point of having one not talk while the other one is talking. So as Jesse was talking and Katie wanted to talk I'd say no Katie, it's not your turn and wait until Jesse's done. Anyway they each gave me their side of the story*	[Think hard. Can you remember anything that he said?] No. I can't remember. [Do you remember what she (Mom) said or did?] Nothing.

Table 15.4 continued

Table 15.4 continued

Younger Sibling's Account	*Parents' Commentary*	*Older Sibling's Account*
When they went, we were going home and Jesse went to her bed. [Mom and Dad said] When we get home, Jesse's, Jesse's going to bed. [Mom and Dad said] Um, um, if you do one more bad thing, you will go to your room.	[Katie Inaccurate] *And I gave them my point that—I told Jesse that Katie probably hit you by accident because you were tackling her and I told Jesse that you shouldn't have tackled Katie in the first place*	[When he talked to you. Did you guys get in trouble?] No. [Did you have to go to your room or anything?] No. We weren't at our house.
	Reconciliation	
She said sorry. We gave a hug to each other. And she gave me big kiss.	[Both Accurate.] *And then I ended up telling the girls to make up—to give each other a hug and say that love you to each of them.* *And Jesse was much more receptive to giving Katie a hug, but Katie was very reluctant, and after a little more persuading Katie finally gave Jesse a hug and after a little more persuading Katie finally did say to Jesse that she loved her.*	[Did dad say anything to you guys at the end? That you had to make up? Or that you had to hug or anything?] Uh, yeah. [And what did you have to say to each other?] Sorry and I love you. Cuz we both said it. Cuz she said the same thing as me.

Note. Mother's comments on the transcript are in brackets and Father's original phone diary version is in italics.

6. Did Katie hit, trip, or kick Jesse? Did she kick her in the head? Or did she take no action whatever once the dandelion changed hands? The parents agree that Katie aggressed, although whether she hit (perhaps accidentally) or possibly pinched Jesse is unclear to them.

7. Did the parents either punish or threaten to punish Jesse, as Katie claims and Jesse explicitly denies? Apparently not. Mom labels Katie's claim as inaccurate.

8. Finally, how reciprocal (as Jesse insists) or one-sided (as Katie would have it) was the reconciliation? Dad describes it as rather one-sided, although Katie does admit and Dad does describe her as taking at least some action, albeit reluctantly, to reestablish amicable relations with her sister.

Of the eight disagreements, the parents are silent with respect to one discrepant claim, support Katie's version five times and Jesse's version twice.

Should the parents' versions be viewed as factual? The adult perspective comes from direct observation of some of the conflict actions, from listening to the children's earlier accounts of the conflict, and possibly from parents' knowledge of the typical behavior patterns of their children. Because the parents are usually neutral observers with respect to sibling conflicts (Martin & H. Ross, 1995; H. Ross et al., 1994), we are tempted to evaluate their versions as especially credible. Yet we should not simply accept the adult accounts as valid, especially when they are based partly on hearsay (the children's reports). It is interesting to note, however, that on average, parents show no systematic bias in assigning culpability. They report equal numbers of transgressions and justifications for older and younger siblings (Smith et al., 1997).

The tendency for children to describe themselves as more aggrieved and less blameworthy than their siblings can be interpreted in two quite different ways. First, perhaps the children are providing us with as complete and accurate an account of the interaction as they can, but their internal representations are biased in favor of their own innocence. Alternatively, the children may edit their portrayals of self and other so as to appear less blameworthy in the eyes of the interviewer than they do in their own eyes. In this case, the representation of events in memory need not be biased in their favor at all. Our data do not provide us with a means of differentiating these two alternatives, but the implications of each are quite interesting.

If children's memories for events are distorted in their favor, then the process of conflict itself may contribute to this bias. During a conflict, children are particularly attentive to their own goals and to the actions of others that prevent the attainment of those goals (Stein & Liwag, 1997; Stein & Miller, 1993; Stein, Trabasso, & Liwag, 1992, 1994; Trabasso & Stein, 1994). Children do display knowledge of the other's goals, but they fail to articulate how their own actions block the other's attainment of valued goals. They do not organize their conflict representations in terms of the other's goals. Instead, the opponent's conflict actions that interfere with what they want form the core of their memory for the events. What is most salient about their own actions is the conditions that impelled them to act as they did. Thus they understand their own actions in terms of the circumstances that led to and justified them, and are more articulate about the justifications for their own transgressions than they are for the transgressions of their siblings (Rueter & Conger, 1995; Smith et al., 1997; Stein & Albro, 1997; Stein, Bernas, & Calicchia, 1997). Within the conflict process itself, then, children are more aware of justifications for their own actions.

Similarly, children are typically more aware of their own emotions. Jesse knows far better than Katie how much her head hurt when it came in contact with Katie's foot. Much information about emotion and motivation is privileged; you know your own feelings directly but must infer those of others (Stein et al., 1994). Some children vividly expressed their irritation with our constant probing to portray what the other thought or felt: "I don't know. I am not her. And I don't have her brains either. And I can't think without her brain." Moreover, negative emotion during conflict should further focus children's attention on their own goals and provide them with less motivation and fewer cognitive resources to view the dispute from the perspective of the opponent (Stein & Levine, 1990; Stein & M. Ross, 1996; Stein et al., 1994).

Children's memories may also be distorted in their own favor because of biases that occur when they later attempt to recall the conflict. People's autobiographical memories are often influenced by their current knowledge and attitudes (M. Ross, 1989). There is a great deal of evidence from the social psychological literature that adults in our culture generally have high opinions of themselves; they evaluate themselves more positively than their friends and peers on a wide range of traits (M. Ross & Wilson, in press; Taylor & Brown, 1988). Particularly pertinent to the current research, university students view themselves as less responsible than their siblings for causing conflicts between the two of them and as better at resolving them (Wilson, Celnar, & M. Ross, 1997). Young children also tend to evaluate themselves more positively than others do and regard their own transgressions as less serious than those committed by other children; they volunteer information about their own successes and others' wrongdoings (den Bak & H. Ross, 1996; Harter, 1998; Smetana, Kelly, & Twentyman, 1984; Stipek, Recchia, & McClintic, 1992). While remembering their sibling conflicts, children in the current study may use their self-knowledge and knowledge of their sibling relationship as cues for retrieving past episodes, for filling in gaps in their memories, and for explaining their previous actions. Children who think very highly of themselves should be unlikely to recall their own misbehavior or to attribute base motives to themselves. They may be much more inclined to remember the wrongdoing of a frequent adversary and to ascribe nefarious motives to a brother or sister. This may help explain why Katie fails to recall aggressing against Jesse. A favorable self-view may also help to explain why neither Katie nor Jesse accepts full responsibility for starting the fighting: The actual cause was perhaps ambiguous and they each view themselves as not being the kind of a child who unilaterally starts fights.

The children's stories of the Dandelion War may also reflect their understanding of how events *ought* to have unfolded. Katie has probably learned from personal experience that a naughty child is sent to bed early. If she were to tell a fictional story about a bad child, she might well incorporate this detail. In erroneously including it in her account of the Dandelion War, she may unwittingly be mixing up episodes—perhaps Jesse has suffered this fate in the past and Katie is simply confused. Katie may also unconsciously be trying to give the story a more fitting conclusion. In Katie's mind, Jesse deserves to be punished, and so she is in Katie's story.

The preceding interpretations of the findings assume that the children generally believe their accounts. It is also possible that children deliberately bias their reports of the dispute in order to manage the impression they were creating for the interviewer. In some ways, this explanation is more intriguing because it assumes that young children possess a store of knowledge that they use to construct flattering versions of conflict. Self-serving biases in conflict reports would require quite detailed and exquisite knowledge about the conflict processes, and of the determinants of culpability and innocence in that domain. The children would need to have knowledge of the moral system that applies within their sibling relationship, including what constitutes a transgression (grabbing the flower from my hands) and an innocent action (picking it up from the ground where it was dropped). The children selectively "don't remember" their own transgressions, report more frequently on the actual misdemeanors of the other, or attribute to the other wrongdoing that did not occur. In our example, Katie directly assigned responsibility for the conflict to her sister. "I didn't start it. Jesse, um Jesse started it first." Moreover, when we divide the transgressions in the full narrative data set into those that are more versus less serious (e.g., physical aggression and property damage are more serious, tattling and bossing less so), we find that it is only the more serious transgressions that children report more frequently for siblings than for self. Other evidence also indicates that children of 4 and 6 years are more likely to inform their parents of their siblings' serious rather than less serious misbehaviors, those whose breaches are more likely to arouse reprimands from parents (den Bak & H. Ross, 1996; H. Ross & den Bak-Lammers, 1998).

More than a system of rules is demonstrated by the biases in the children's accounts, if such biases reflect intentional deceit. Children also know how to exaggerate the culpability of others and how to justify the actions that they themselves take within conflict. The range of justifications is quite broad. In the sisters' accounts of the Dandelion War, justifications include statements that minimize the damage caused by own actions, that attribute altruistic motives to the self, that claim rights such as possession or ownership, that cite the provocative actions of others, that note that negative actions were reciprocal, and that invoke external, uncontrolled causes of own actions. In the full data set, children also described their actions as necessary to avert a greater harm, unintentional, and occurring with the consent of the victim. In addition, in discussing their oppenent's actions, the children emphasize negative consequences and malicious intent; children are less likely to note such factors while describing their own misdeeds.

To cast themselves as more innocent and the other as more blameworthy, the children argue their cases based on a highly nuanced system of moral rules and responsibilities. If the children are indeed telling the interviewer a different story about the conflict than they are telling themselves, then they must have quite explicit knowledge of the rules that operate within their families and of the factors that mitigate the culpability of those who violate the rules. From the third year, children do use such knowledge to argue logically and appropriately with one another to advance their own goals within disputes (Albro, 1996; Dunn & Munn, 1987; H. Ross, 1996; Tesla & Dunn, 1992). In telling a naive questioner about a dispute, the children must also carefully craft a narrative that convincingly portrays their edited version of events. We find that skill more clearly illustrated in the stories of our 6-year-old narrators than in those of the 4-year-olds, but clearly present in the tales of children of both ages.

If young children's reports of the conflicts are, at least in part, an exercise in impression management, one can ask how children acquire these self-presentational skills. A likely answer is that they get a lot of practice describing conflicts with their siblings to their parents (den Bak & H. Ross, 1996; H. Ross & den Bak-Lammers, 1998). In his description of the Dandelion War, the father notes:

> First I asked Jesse what happened so Jesse told me the story and then I asked Katie what happened and Katie told me her story. And it was difficult for

> them to not talk while the other one's talking. I tried to make a point of having one not talking while the other one was talking. So as Jesse was talking and Katie wanted to talk I'd say "no Katie, it's not your turn and wait until Jesse's done." Anyway they each gave me their side of the story.

In such interactions, parents socialize their children to talk about their conflicts in certain ways. In this example, they are required to take turns and not to interrupt each other. Note also the assumption that each child will tell *her side* of the story. In the course of such family discussions, children may learn that they are rewarded for portraying their actions favorably and for defending themselves against the charges of their siblings. They may learn to be less concerned about truth and more concerned about effectively arguing their cases.

The occurrence of these family discussions may muddle our distinction between memorial and self-presentational biases. The conflict account that children present to the interviewer is sometimes not the first time they have told the story of that dispute. Katie and Jesse may tell us the same stories that they told their parents because their memory of what they told their parents is vivid and has altered their current recall of the dispute. People sometimes come to believe their self-presentations; their biased portrayals affect their memory of events (Higgins & Rholes, 1978). These reflections lead us to wonder whether the content of children's stories of disputes changes as a function of tattling to their parents and whether children truly come to believe their biased portrayals. A biased portrayal of their own innocence might have more practical implications for the children at home than in our study. However, if "saying" can lead to "believing," then their memory of the conflict may be influenced by their recollections of past narratives, as well as by their recall of the actual events.

Note that our interpretations of the data have focused on understanding the conflict narratives, rather than on the origins of the conflict per se. An interest in the bases of the conflict would lead one to inquire, for example, why the children chose to fight over the fate of a single dandelion. Dandelions, like mice and rabbits, tend to be abundant. We suspect that Katie and Jesse could have walked to opposite ends of the field and collected enough flowers to satisfy both of their needs. The dandelions appeared to accrue special value by being in the possession of one or the other sister (Hay & H. Ross, 1982). Such analyses go beyond the goals of our study, however. We examine children's perspectives on conflict, not why they fight. Nonetheless, we strongly believe that an understanding of the former will contribute to an understanding of the latter.

It is clear that our data do not allow us unequivocally to attribute the biased narratives to either memory processes that reflect the perspectives of the children during and after the conflict or the impressions that the children are deliberately trying to create either for their parents or for the interviewers after the fact. However, it would appear that our data force us to consider either one or the other (or even both) as likely possibilities, and that does narrow the range of conclusions considerably. It is not the case that memory for conflict features the children's own actions more so than those of the other, or that 4- and 6-year-old children know little about the structure of their sibling disputes. The current challenge that we face is to identify the processes that underlie the biases children display in their reports of their disputes. We expect that the most profitable research approach will combine narrative analysis with experimental manipulations of children's motivation for reporting conflicts. If the biases reflect children's memories of the events, then the biases will persist when children strive for maximum accuracy. If the biases reflect efforts at impression management, then they should be exaggerated when children's reports have consequences that matter to them.

ACKNOWLEDGMENTS

We thank the families who participated in this study, especially Jesse, Katie, Daniel, and their parents. The research was supported by grants from the Social Sciences and Humanities Research Council of Canada.

REFERENCES

Albro, E. R. (1996). *The role of liking in children's memories and appraisals of peer conflicts.* Unpublished doctoral dissertation, University of Chicago, Chicago.

Baumeister, R. E., Stillwell, A., & Wotman, S. R. (1990). Victim and perpetrator accounts of interpersonal conflict: Autobiographical narratives about anger. *Journal of Personality and Social Psychology, 59,* 994–1005.

den Bak, I. M., & Ross, H. S. (1996). I'm telling! The content, context, and consequences of children's tattling on their siblings. *Social Development, 5,* 292–309.

Dunn, J., & Munn, P. (1987). The development of justification in disputes with mother and sibling. *Developmental Psychology, 23,* 791–798.

Harter, S. (1996). The development of self-representations. In N. Eisenberg, (Ed.), W. Damon (Series Ed.), *Handbook of child psychology: Vol. 3. Social, emotional, and personality development* (5th ed., pp. 553–617). New York: Wiley.

Hay, D. F., & Ross, H. S. (1982). The social nature of early conflict. *Child Development, 53,* 105–113.

Higgins, E. T., & Rholes, W. S. (1978). "Saying is believing": Effects of message modification on memory and liking for the person described. *Journal of Experimental Social Psychology, 14,* 363–378.

Martin, J. L., & Ross, H. S. (1995). The development of aggression within sibling conflict. *Journal of Early Education and Development, 6,* 335–358.

Ross, H. S. (1996). Negotiating principles of entitlement in sibling property disputes. *Developmental Psychology, 32,* 90–101.

Ross, H. S., & den Bak-Lammers, I. M. (1998). Consistency and change in children's tattling on their siblings: Children's perspectives on the moral rules and procedures of family life. *Social Development, 7,* 275–300.

Ross, H. S., Filyer, R. E., Lollis, S. P., Perlman, M., & Martin, J. (1994). Administering justice in the family. *Journal of Family Psychology, 8,* 254–273.

Ross, M. (1989). The relation of implicit theories to the construction of personal histories. *Psychological Review, 96,* 341–357.

Ross, M., & Wilson, A. (in press). Construing and appraising past selves. In D. Schacter (Ed.), *Memory and belief.* Cambridge, MA: Harvard University Press.

Rueter, M. A., & Conger, R. D. (1995). Interaction style, problem-solving behavior, and family problem-solving effectiveness. *Child Development, 66,* 98–115.

Shantz, C. U. (1993). Children's conflicts: Representations and lessons learned. In R. R. Cocking & K. A. Renninger (Eds.), *The development and meaning of psychological distance* (pp. 185–202). Hillsdale, NJ: Lawrence Erlbaum Associates.

Smetana, J. G., Kelly, M., & Twentyman, C. T. (1984). Abused, neglected and nonmaltreated children's conceptions of moral and social-conventional transgressions. *Child Development, 55,* 277–287.

Smith, M., Wilson, A., Ross, H., & Ross, M. (1997, April). *"I didn't mean to." Children's accounts of their conflicts with their siblings.* Paper presented at the meeting of the Society for Research in Child Development, Washington, DC.

Stein, N. L., & Albro, E. R. (1997). *Children's understanding of conflict: Evidence from past memories and on-line negotiated resolutions.* Paper presented at the meeting of the Society for Research in Child Development, Washington, DC.

Stein, N. L., Bernas, R., & Calicchia, D. (1997). Conflict talk: Understanding and resolving arguments. In T. Givon (Ed.), *Typological studies in language: conversational analysis*(pp. 233–267). Amsterdam: John Benjamins.

Stein, N. L., & Levine, L. (1990). Making sense out of emotional experience: The representation and use of goal-directed knowledge. In N. L. Stein, B. Leventhal, & T. Trabasso (Eds.), *Psychological and biological approaches to emotion* (pp. 45–73). Hilsdale, NJ: Lawrence Erlbaum Associates.

Stein, N. L., & Liwag, M. D. (1997). A goal-appraisal process approach to understanding and remembering emotional events. In P. van den Broek, P. Bauer, & T. Bourg (Eds.), *Developmental spans in event comprehension and representation* (pp. 199–235). Mahwah, NJ: Lawrence Erlbaum Associates.

Stein, N. L., & Miller, C. A. (1993). A theory of argumentative understanding: Relationships among position preference, judgments of goodness, memory, and reasoning. *Argumentation, 7,* 183–204.

Stein, N. L., & Ross, M. (1996). *Discussing conflicts: Similarities and differences in husband's and wife's description of the same conflicts.* Paper presented at the meeting of the Psychonomic Society, Chicago.

Stein, N. L., Trabasso, T., & Liwag, M. (1992). The representation and organization of emotional experience. Unfolding the emotional episode. In M. Lewis & J. Haviland (Eds.), *Handbook of emotion* (pp. 279–300). New York: Guilford.

Stein, N. L., Trabasso, T., & Liwag, M. (1994). The Rashomon phenomenon: Personal frames and future oriented appraisals in memory for emotional events. In M. Haith, J. B. Benson, R. J. Roberts, & B. F. Pennington (Eds.), *The development of future-oriented processes* (pp. 409–435). Chicago: University of Chicago Press.

Stipek, D., Recchia, S., & McClintic, S. (1992). Self-evaluation in young children. *Monographs of the Society for Research in Child Development, 57*(Serial No. 226).

Taylor, S. E., & Brown, J. D. (1988). Illusion and well-being: A social psychological perspective on mental health. *Psychological Bulletin, 103,* 193–210.

Tesla, C., & Dunn, J. (1992). Getting along or getting your own way: The development of young children's use of argument in conflicts with mother and sibling. *Social Development, 1,* 107–121.

Trabasso, T., & Stein, N. L. (1994). Using goal-plan knowledge to merge the past with the present and the future in narrating events on line. In M. Haith, J. Benson, R. Roberts Jr., & B. Pennington (Eds.), *The development of future-oriented processes* (pp. 323–349). Chicago: University of Chicago Press.

Trabasso, T., Stein, N. L., & Johnson, L. R. (1981). Children's knowledge of events: A causal analysis of story structure. In G. Bower (Ed.), *Learning and motivation* (Vol. 15, pp. 237–281). New York: Academic Press.

Wilson, A. E., Celnar, C., & Ross, M. (1997, August). *Siblings' self and other perceptions of past and present conflict.* Poster presented at the American Psychological Association Conference, Chicago.

16

Predicting Psychological Well-Being From Beliefs and Goal-Appraisal Processes During the Experience of Emotional Events

Nancy L. Stein
Rachael A. Sheldrick
Sara C. Broaders
University of Chicago

During the last 10 years, we have examined the significance of the appraisal process during children's and adults' recall of emotional, traumatic, and stressful events. Our studies provide an assessment of the personal meaning and memory for emotional events, and the ways in which these events are encoded and recalled. Because the emotional appraisal process is not well understood or well documented, we have chosen to focus on describing the process of appraising and understanding events that specifically induce emotional reactions. We focus on the ways in which language, talk, and planning can be used to describe the types of appraisals that occur during the experience of an emotion, the ways in which memories for emotion are organized, and the predictive validity of appraisals in relationship to psychological well-being in both children and adults.

Using memories of emotional events to predict psychological well-being is not a new enterprise. As Kihlstrom (in press-a, in press-b) demonstrated, Freud had most of his ingenious insights into psychological well-being by studying the past memories of his patients. Freud believed that the ways in which people structured their memories would directly predict current and future states of psychological lucidity and mental health. Indeed, many current investigations of mental health (Beck, 1988; Gottman, 1994; Michenbaum, 1994; Traue & Pennebaker, 1994) use narrative interviews or written reports of conversations between partners (Ross & Holmberg, 1990; Stein & M. Ross, 1996) to predict the success of a relationship and personal states of psychological well-being.

In the studies we describe in this chapter, children, adolescents, and adults have been asked to talk about emotional, stressful, or conflictual events that have had an impact on their everyday functioning. In one set of studies (Levine & Stein, 1999; Stein & Albro, 1997), we examined the nature and impact of conflicts between 3- and 5-year-old preschool children and their parents. In a second study (Stein, Ross, Sheldrick, & Fergusson 1999), we examined the impact that traumatic events had on the psychological well-being of adolescents. In a third set of studies (Stein, Trabasso, & Lichtenthal, 1999), we described the nature of stressful events that male caregivers experienced while caring for a partner who had AIDS and eventually died of the disease. We have also described the bereavement process (Stein, Folkman, Trabasso, & Richards, 1997) and the process that caregivers go through as they recover from the death of their partner (Stein & Chae, 1999).

The three sets of studies are strikingly different. Each focuses on a different age group; each deals with different types of intimate or familial relationships; and each has a different set of themes that characterize the experiences of the participating children and adults. The common element that bridges the three studies, however, is a focus on the representation of emotional, stressful, or traumatic experiences and the ways in which appraisals of these experiences have affected psychological well-being. Our goal is to document the commonalties and differences that exist across development, the early emergence of young children's ability to understand and interpret emotional events in the same ways that adults do, and children's rapid development of the same types of causal reasoning, desires, and planning skills that adults use.

The reasons for pursuing issues of similarities as well as differences in the development of emotional understanding are related to our goals of describing the learning and coping strategies that result in positive well-being and increased understanding of other people. We have found that talk about emotional states, strategies used for coping with the impact of emotion, beliefs held about the ability to survive or tolerate aversive consequences of stressful events, and the ability to generate new goals in the face of loss show more similarities than differences across the life span (see also Trabasso & Ozyureck, 1997). Individual variation within an age group is far greater than any developmental differences we have found in emotional understanding.

We have also found that young children, compared to adults, are often more accurate, more flexible, and more adaptive in family conflict situations, especially those that require negotiation and compromise. Children do not have as much access to overt power as their parents do, and so they quickly learn "survival" tactics to interact with other family members, especially in relation to getting what they want and protecting themselves from harm, if at all possible. In some situations, children know as much about their parents as their parents know about them. Although some developmentalists believe that children become less egocentric, more socially astute, and more adept at social negotiation as they get older, we, as well as other developmental researchers, have found that the growth of social awareness and ability does not proceed according to any working stage model of development (Trabasso & Stein, 1994, 1997). Rather, personal preferences, goals, and personal tendencies that children and adults possess, such as the desire to be dominant, to have close personal relationships, to be recognized as the best, to be the most powerful, to help other people, to be alone, or to belong to a group, have proven to be more predictive of strategies that are adopted in emotional situations than are the ages of the participants.

Developmental differences do emerge in emotion narratives, but they have little to do with the basic types of appraisals that are made about an emotional experience. The two developmental differences we and others (Albro & Stein, 1997; Stein, 1984, 1988; Goldman, 1982; Levine & Stein, 1999; Stein & Trabasso, 1982, 1989) have found is that older children and adults have more knowledge about obstacles that stand in the paths of goal attainment, and they have more knowledge about plans that will accomplish their goals (see Haith, 1994). Thus, when a younger and older child get into a fight, it is the older sibling who is more often the winner and who often displays more cunning and deception than the younger child (H. Ross, 1996). This cunning, however, is not necessarily linked to the older child's better understanding of what the younger sibling is thinking, nor is the cunning related to increase in social skill. Rather, the older child knows much more about how to get what he or she wants, is faster at formulating a plan that thwarts the younger child, and is more persistent in pursuing his or her own goals (Stein & Bernas, 1999).

The ways in which children and adults organize and structure knowledge about emotional events has been shown to be quite similar (Levine, Stein, & Liwag, 1999; Stein & Levine, 1989). Children and adults both use conscious and unconscious processes to understand, evaluate, and respond to events that evoke emotion. From the very beginning, emotional understanding is goal and preference based. Even the youngest children who experience and express what we consider to be the "basic" emotions (e.g., happiness, sadness, fear, anger, and contempt) make inferences about the harms and benefits of emotion-laden situations. Very young children also engage in planful action to attain pleasurable states or to eliminate aversive states. Thus, the mental structures that govern and regulate emotional understanding are in place well before children begin to talk. We argue that the types of appraisals that children carry out in emotion situations are similar to those that adults carry out.

To document the early emergence of emotional understanding, we present a model that operates from at least 2 years of age, when children begin to talk coherently and communicate their understanding of emotional events. We describe how these young children evaluate events, goals, plans, and actions during the retrospective recall of emotional events. We also discuss studies in which young children narrate online during the experience of emotion.

The early emergence of emotional understanding does not negate the existence of developmental differences in the depth of understanding of emotion-eliciting events (Stern, 1985, 1992). Typically, adults have been exposed to more situations than children, and as a result have broader knowledge of the causes and consequences of emotion-eliciting events than do young children. The question we pose, however, is whether these developmental changes influence the basic organization of emotional thinking, or whether they influence the specific types of appraisals made, the specific coping strategies used, and the speed with which retrieval of information occurs.

A MODEL OF EMOTIONAL EXPERIENCE AND UNDERSTANDING

Four dimensions characterize the process of understanding that accompanies emotional experience. First, the evocation of emotion always signals that some type of change has been perceived in a personally significant goal. That is, children and adults who experience an emotion have perceived some type of change in one of their goals such that the goal has been attained, blocked, or threatened (Folkman & Stein, 1997; Stein & Levine, 1987, 1989, 1990; Stein, Trabasso, & Liwag, 1992).

Second, a one-to-one correspondence between a given event and a specific emotion does not exist. The experience of emotion depends on the prior knowledge, beliefs, and appraisals used to assess the status of personally significant goals. The appraisals made about the causes and consequences of an event, rather than the event itself, determine whether and which emotion will be experienced. The lack of a direct correspondence between an event and an emotion holds even for young children (Stein & Levine, 1989; Stein, Liwag, & Wade, 1995).

As an example, Sroufe (1979) noted that very young infants can have a variety of reactions to the game of peek-a-boo. At one time, an infant may respond to the game with a startle response, followed by a fearful expression and crying. At another time, the same infant may display gleeful laughter when engaged in the game of peek-a-boo. The critical factors believed to regulate the experience of an emotion are the initial psychological state of the infant, in terms of the goals and beliefs operating at the onset of a specific event, and the inferences made about the consequences the event will have for a valued goal state.

The involuntary nature of emotional experience is the third dimension that characterizes the process of understanding. We do not plan to experience an emotion. We can anticipate how we might react when an emotion-laden event occurs, and we can even visualize how we might respond to a specific event (as therapists are wont to have their patients do in practicing emotion regulation). However, emotions occur involuntarily and unexpectedly, in response to novel aspects of an event or situation. The presence of information that violates prior expectations is a critical component of emotional experience (e.g., G. Mandler, 1984; Stein & Levine, 1987).

As an example, when children and adults are presented with the same event over time, they may experience an emotion the second time because they perceive a new element in the event that was not encoded during the first presentation. If children and adults fail to notice a novel aspect of an event, however, they do not experience the same intensity of emotion, and sometimes they report not experiencing any emotion (Stein & Trabasso, 1989). For example, think of a favorite movie or song. Initially, seeing the movie or hearing the song may produce a strong emotional reaction. Over time, however, with frequent exposure, the movie or song is no longer novel, and does not provoke an emotional response any longer. Instead, one may actually become bored with it because there is no novelty remaining.

The fourth dimension that characterizes emotional understanding is the causal reasoning, goal-appraisal, and planning processes that operate continually throughout the experience of an emotion. Current theories of cognition and emotion have all incorporated the importance of goal-appraisal processes (Frijda, 1995; Oatley & Johnson-Laird, 1987; Roseman, Antoniou, & Jose, 1996; Smith & Lazarus, 1993). Frijda used the word *concerns* to illustrate the focus on goals; Oatley and Johnson-Laird examined goals that differentiate among the emotions; Smith and Lazarus spoke to the importance of linking appraisals to the goals of well-being. We focus on the goal-appraisal processes that are carried out throughout the experience of emotion, and we describe the ways in which thinking and appraisal unfold during emotional experience (Folkman & Stein, 1997; Levine, 1996; Stein, Folkman, Trabasso, & Richards, 1997; Stein & Levine, 1987, 1990; Stein & Trabasso, 1999; Stein, Trabasso, & Liwag, 1992, 1994).

The model that we present in this chapter is distinct from other models in the following ways. We address issues related to the *development* of emotional understanding. Thus, we examine understanding in very young children (Levine, 1995; Levine & Stein, 1997; Liwag & Stein, 1995; Stein & Levine, 1989; Stein & Liwag, 1997; Stein et al., 1994: Trabasso, Stein, & Johnson, 1981) and in adults (Levine, 1996; Stein, Folkman, et al., 1997; Stein & Levine, 1989; Stein et al., 1994). Rather than proceed with the assumption that children are fundamentally different thinkers than adults, we use our theory of the understanding process to examine the similarities and differences between children's and adults' thinking. In this way, we are able to speak to the universals associated with emotional understanding as well as to the differences that may occur.

We also describe appraisal, thinking, and planning, as these processes occur *online* during the experience of an emotion (Stein, Bernas, & Calicchia, 1997; Stein & Boyce, 1995; Trabasso & Ozyureck, 1997) or *retrospectively,* as people engage in recalling events that are emotionally meaningful (Levine, 1996, in press; Levine & Stein, 1997; Liwag & Stein, 1995; Stein & Liwag, 1997; Stein et al., 1995; Stein, Wade, & Liwag, 1997). The emphasis on content, causal sequencing, and function allows us to describe the "unfolding" of an emotion episode.

In our analysis of the nonverbal dimensions of emotion, we focus on the expressive systems of the face, voice, and hands as well as the sequence of nonverbal actions that children and adults produce during the "narration" of emotional experience (Stein & Boyce, 1995).

Finally, we have examined the accuracy of memory for emotional events, the ways in which memory affects subsequent states of psychological well-being, and the ways in which emotions affect thinking and strategic interaction during conflicts (Albro & Stein, 1997; Folkman & Stein, 1997; Levine, in press; Levine & Burgess, 1995; Stein, Bernas, & Calicchia, 1997;

Stein, Folkman, et al., 1997; Stein & Trabasso, 1992). We also characterize individual differences in emotional experience and describe the impact these differences have on memory for emotional events (Stein & Boyce, 1995).

THE EMERGENCE OF EMOTIONAL UNDERSTANDING

Given that thinking and appraisal are integral to emotional experience, what can we say about their development? The experience and expression of emotion change significantly during the first 6 months of life. At birth, the infant's affective repertoire includes a set of behaviors for responding to different types and intensities of stimulation. Some of these behaviors are reflexive in nature (e.g., the startle, orienting, blinking, and sucking reflexes) and are followed by general affective responses like distress (e.g., volatile activity and crying), quieting (e.g., the absence of volatile activity), and pleasure.

Some researchers take a reflexive response, coupled with a general affective response, such as distress or quieting, to be indicative of emotion (LeDoux, 1996). In our theory (Stein & Levine, 1987, 1989; Stein et al., 1992), these general affective responses are not defined as an emotion. The experience of emotion requires that infants be able to represent goal states and the changes that occur in these goal states. The ability to represent a state that is not yet present (e.g., a goal state) gradually emerges between the ages of 2 to 6 months (Sroufe, 1979; Stenberg & Campos, 1990). The ability to exhibit a preference for one state versus another also emerges at approximately the same time. It is when infants acquire the capability to represent an absent goal state, prefer one state over another, and invoke action in the service of getting from one state to another, that they begin to experience and express emotion. An important source of evidence for this development is that generalized negative reactions such as the distress expression gradually give way to discrete facial expressions of anger, sadness, and fear that are accompanied by actions appropriate to these emotional states (e.g., Izard, Hemsbree, & Huebner, 1987; Stenberg & Campos, 1990).

LeDoux's (1996) finding of a direct connection between the sensory thalamus and the amygdala is not evidence that emotional responses are "hard-wired" or that emotional responses bypass the cortex. The level of behavior LeDoux described is considered "reflexive" by many, and his data pertain to the behavior of rats. What LeDoux defined as "emotional" we define as "affective." The critical data that LeDoux needed in humans are those from decerebrate infants, and the few inquiries that exist indicate that discrete expression of emotions never emerges in these infants (Kolb & Taylor, 1990).

The fact that newborn infants express pleasure and distress does not mean that they prefer or desire to shift from one state to another. Having preferences requires the ability to represent, remember, compare, and choose between two different states, and to express a desire to orient more toward one state than toward another. Although newborns display pleasure or distress in direct response to different events, they have yet to achieve the capacity to represent a state different from the one currently directing their behavior. The time between birth and 6 months of age is believed to be essential for acquiring the prerequisite knowledge necessary to experience and express emotion as we define it. Several milestones can be observed. The infant's ability to habituate and to remember different classes of events increases in breadth. Habituation signals that infants can form predictable representations of an event and that they can detect discrepant information in the event (or dishabituate) when novelty is introduced (Alessandri, Sullivan, & Lewis, 1990; Lewis, Alessandri, & Sullivan, 1990).

SIMILARITIES IN EMOTIONAL UNDERSTANDING ACROSS THE LIFE SPAN

To illustrate the similarities and differences in emotional understanding across the life span, we use the following evidence: (a) the events and appraisals that precipitate happiness, sadness, anger, and fear; (b) the desires and plans activated in response to the experience of each emotion; and (c) the relationship between positive and negative appraisals of the emotion event, coping strategies used in response to the goal change, and psychological well-being, as measured by depressive mood and positive morale.

Events That Precipitate Emotion

Tables 16.1 and 16.2 include the types of events that children and parents nominated as precipitators of happiness, anger, sadness, and fear in themselves. Children in our studies were from 2½ to 5 years of age. Their parents were between the ages of 37 and 53 years and were selected primarily from the middle and lower middle class. Approximately 65% were White, 25%

were African American, and the remainder were from mixed ethnic backgrounds.

Tables 16.1 and 16.2 display the most prototypic events that parents and children nominated as precipitating each emotion. The two sets of nominations are highly similar to each other, especially when nominations are compared *across the four different emotions.* As an example, in both samples, the events nominated for happiness do not overlap with the events nominated for the three negative emotions. Happiness is about getting what you want, and sadness is about losing it. Similarly, events nominated for anger and fear do not overlap. Anger occurs when someone breaks a promise, takes away a valued possession, intrudes upon the participant, and so on. Fear occurs when the individuals perceive a threat to themselves or others, when they perceive another animate being, such as a ghost, monster, robber, and the like, or when they are exposed to noxious, aversive sensory stimuli that they cannot escape.

TABLE 16.1

The Prototypic Events Generated by Parents and by Children That Elicit Happiness and Sadness

Emotion	*Event Categories*	*Percentage Parents*	*Percentage Children*
Happiness	Engages in/masters desirable activities	22	15
	Gets desirable objects	30	25
	Participates in family activities	14	3
Sadness	Separated from significant others (death of other)4220	42	20
	Is physically harmed	0	23
	Loss of favored object	15	35

TABLE 16.2

The Prototypic Events Generated by Parents and by Children That Elicit Anger and Fear

Emotion	*Event Categories*	*Percentage Parents*	*Percentage Children*
Anger	A family member breaks a promise	30	20
	Possessions are taken away or destroyed	14	2
	Forced to do something	14	3
	Physically harmed	0	20
	Intruded upon	30	31
Fear	Perceives a threat to self or others	53	12
	Experiences undesirable sensory stimuli	22	15
	Emotions are elicited by animate beings	0	30

Figure 16.1 contains the statistical distance scale that was used to measure the amount of overlap between the nominations for each emotion. The scores are correlative in nature and are based on first illustrating the most amount of overlap between the four groups to the least amount of overlap. The greater the distance score, the less the overlap in nominations. Anger and sadness have the highest degree of overlap in terms of events that precipitate each emotion, and therefore the smallest distance score differentiates them. The small distance that separates the two emotions is present in both children's and adults' nominations of anger and sadness.

We (Stein & Levine, 1989, 1999) explained this overlap by the fact that a single event can precipitate more than one emotion, depending on which dimension of the event is being considered. When children or adults lose a valuable object or relationship, they are sad when attending to the impact of the loss in terms of maintaining goals that they value. They also attend to who caused the loss of an object or relationship. If someone intentionally caused the loss, they also experience anger directed toward the perpetrator. Thus, events that evoke anger and sadness are often the same. Sadness is focused on the impact of the loss; anger is focused on the cause of the loss (Stein & Levine, 1987; Stein et al., 1992).

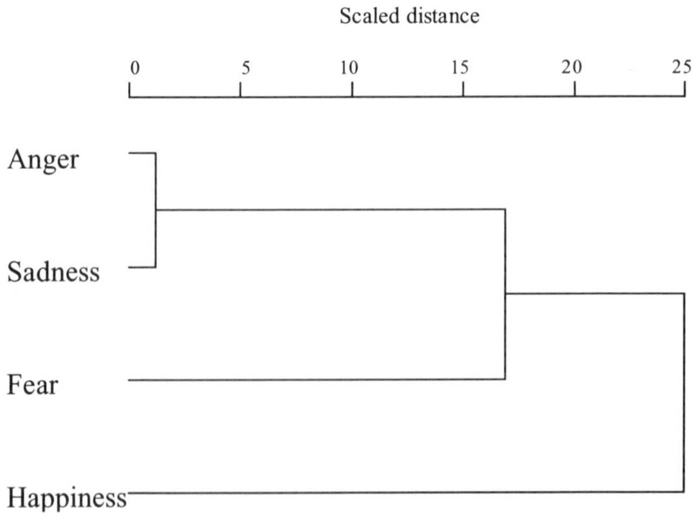

FIG. 16.1. Cluster analysis of event categories. The smaller the scale score, the more overlap in the types of events nominated for each emotion

The distance scale reveals little overlap in events that provoke fear in comparison to those that provoke anger and sadness, and clearly there is even less overlap in the events that provoke positive versus negative emotions. Thus, events that precipitate happiness and fear have little in common with each other and with anger and sadness. Happiness occurs when people succeed in getting what they want or in avoiding a situation that is aversive or harmful to them. Fear occurs when a threat is perceived to a goal that is valued. When people experience fear, they are quite certain that something bad is going to happen, but it has not yet happened. The reason that some overlap exists for fear with both happiness and sadness is quite logical. Happiness and fear may overlap because the same event may have led to attainment of a valued goal (happiness) while also leading to some uncertainty that the goal can be maintained in the long term (fear). The distance scale shows that although this is not a common occurrence, it can occasionally occur. Fear and sadness overlap slightly for a similar reason. It is sometimes a small step from the possibility of losing a goal (fear) to actually losing it (sadness).

Within the class of events that evoke a particular emotion, differences *do exist* in the frequency with which parents and children nominate events. For example, parents associate sadness with the loss of a relationship more frequently than children do, whereas children associate sadness more frequently with losing a favorite object. These differences are explainable by the age and experience of children and adults. Many children have not suffered the number of losses that adults have, whereas most of them have experienced the loss of a favored object (toy or game). Indeed, children are very quick to point out all of the objects that they have lost. If preschool children are asked about their feelings *should they* experience the loss of a relationship, however, almost all respond with the emotion of sadness (Stein & Jewett, 1986). Thus, children are clearly able to conceive of feeling sad in response to the loss of a relationship when they encounter hypothetical situations. However, they rarely nominate these types of events as actually provoking sadness in everyday interaction.

The same can be said about adults experiencing sadness as the result of loss of a favored object. Adults clearly respond with sadness when a hypothetical scenario is presented where a protagonist has lost a valued object, but in reality, when describing everyday events, the most traumatic events are typically those that involve relationship loss. The same is true of adolescents who nominate events that have been the most traumatic for them. Over 50% of all traumatic events concerned the loss of or change in a relationship.

Variation in the frequency with which events are nominated, however, could be due to factors that have little to do with development. Husbands and wives (Stein & Ross, 1996) differ as much as parents and children, and two siblings could differ as much as parents and children. The question of prototypic frequency, as representing particular themes for each age group, will need to be pursued in future studies. However, experience, gender, and culture appear to be factors that affect prototypic frequency of nomination for each emotion as much or more than age does.

Goal-Outcome Appraisals as Reasons for Emotions

The next set of data pertains to the reasons children and adults gave for experiencing happiness, anger, sadness and fear. Reasons for emotions focus mostly on goal-outcome states. Reasons are given spontaneously during the recall of an emotional or stressful event or in response to "why" probe questions focusing on reasons for experiencing each particular emotion. As the data indicate in Table 16.3, the goal-outcome states reported for happiness are *distinct* from goal-outcome states reported for the three negative emotions. *Getting something* that a person wanted (e.g., a desired toy or favorite food, a desired vacation, etc.) was most frequently associated with the experience of happiness.

Goal-outcome states associated with anger and sadness were limited to desired states that were *not attained* (wanting something but failing to get it) and undesired states that *had to be tolerated* (having to put up with an aversive state). Although both of these goal-outcome combinations were generated for anger and sadness, aversive (undesired) states were associated with anger more than with sadness. Conversely, children and adults asso-

TABLE 16.3

Percentage of Children and Adults Mentioning Different Goal/Outcome Appraisals as Reasons for the Experience of Each Emotion

	Emotions			
Goal/Outcome Appraisals	*Happy*	*Angry*	*Sad*	*Afraid*
Want-Have It	76	0	2	29
Not Want-Not Have It	15	43	65	0
Want-Not Have It	0	0	0	0
Not want-Have It	7	57	33	0
Want-Have It (Uncertain)	2	0	0	71

ciated loss of a desired state with sadness more than with anger. Fear was associated with the mention of goals that had not yet been affected, but all participants were positive that changes would rapidly occur in these states. Thus, fear was most frequently mentioned as a threat—something that had not yet happened, but would certainly happen in the near future (Liwag & Stein, 1995; Stein & Liwag, 1997; Stein, Liwag, & Wade, 1995).

These results replicate and support those reported by Stein and Levine (1989), who examined children's and adults' causal understanding of anger, sadness, and happiness. Stein and Levine discovered that both children and adults were more likely to respond to an aversive state with anger than with sadness, whether or not an animate agent caused the aversive state.

Unlike nominations for each emotion, which showed some variation due to development, the core reasons for experiencing each emotion show no such variation whatsoever. This lack of developmental difference is important because it indicates the many constraints imposed on thinking of emotion and goal-directed action. Our theory (Stein & Levine, 1987; Stein et al., 1992) states that the primary reason that any emotion is evoked is that events occur in an unexpected fashion and directly affect the status of a valued goal. The ways in which events affect goals are to either block them, facilitate their attainment, or threaten their existence. Thus, in any explanation for emotion, we would expect to find continual reference to the types of state changes that have occurred in valued goals. This is exactly what happened when adults and children were asked for their explanations for emotion states.

Our theory of emotional understanding also states that emotions occur because changes in goal states are unexpected and violate existing expectations about what should have occurred rather than what did occur. We have found that people do talk about violations of expectation in explaining their emotional reactions, but this talk is reserved for those events that we would classify as traumatic or life altering in some way. Two of our studies (Stein, Folkman, et al., 1997; Stein & M. Ross, 1996) on traumatic experiences have indicated that when events challenge core beliefs about the world, the reasons given for emotional involvement are filled with talk about violations of expectations.

As an example, when adolescents were asked to talk about some of their most traumatic experiences, these are the types of expressions they reported when providing explanations for emotions: I don't think it's fair of God, destiny, fate, or whatever; Let death take old dying geriatric patients, not the youthful soul that I love; How could anybody be so cruel, I can't believe it happened; How could people be so barbaric; I came to the crushing realization that Jason was gone, I couldn't believe it. If Jason just up and died, so could I; I thought my parents were gone forever; and I thought no one would ever help me. These comments are made when events result in death, in serious injury, in loss of a parent or family member, in loss of a job, in loss of academic standing, and so forth.

In expressing emotion on an everyday basis, although the emotional reactions are often volatile and intense, attention tends to be more focused on solutions that can be generated to solve the problem or deal with the emotional reaction itself. Traumatic events, however, do not have a quick solution and often require the complete abandonment of a *system* of meaningful goals. Rather than one goal being jeopardized, a whole set of related goals is threatened. The more goals that are affected by an event, the harder it is to generate a plan to solve the problem caused by a precipitating event. In these circumstances, people begin to consciously question the validity of the beliefs they have held about the world, and this questioning leads them to actively express and examine the specific violations that affected their beliefs.

Strategies for Coping With Each Emotion

Just as specific emotions are associated with specific precipitating events and reasons, so are they associated with specific desires and strategies for coping with the changes that have occurred to valued goals (see Stein et al., 1992; Stein, Trabasso, & Albro, 1997 for an overview). Table 16.4 contains the strategies generated for coping with the four different emotions. Because we have found no developmental differences in the general categories of desires and plans generated to cope with specific emotions, the data from children and adults were collapsed.

TABLE 16.4

Percentage of Children and Adults Mentioning Different Strategie for Coping With the Changes That Occurred With Each Emotion

	Emotion			
Wishes	*Angry*	*Sad*	*Afraid*	*Happy*
Maintain goal	0	0	0	37
Attain new goal	0	0	0	47
Reinstate original goal	40	36	8	16
Substitute new goal	20	46	0	0
Abandon original goal	0	2	0	0
Revenge	40	16	0	0
Prevent undesirable state	0	0	92	0

The most important finding across all age groups is that the prototypic strategies generated for each emotion do not overlap. For happiness, the two most frequent strategies are: attempting to maintain the goal that was attained, or pursuing a new goal given that the current goal was satisfied. The language of happiness was unique and never occurred in narratives about negative emotions.

Two primary coping strategies were associated with anger: reinstating the failed goal and revenge. When the goal under consideration could not be reinstated, almost all participants oriented toward revenge. Sadness resulted in the use of two strategies: substituting a new goal and attempting to reinstate the failed goal. What is unusual about the reinstatement coping strategy associated with sadness is that both children and adults are fully aware that the goal they want is simply not attainable. However, some refuse to give up thinking about these lost goals. Rather, they spend a great deal of time either ruminating about the intense pleasure they received when the goal was attainable, or engaging in fantasy, pretending that the goal is still feasible. For the moment, or however long the fantasy lasts, the goal is theirs again. This rumination, however, is directly related to subsequent feelings of depression, as our mental health outcome analyses indicate.

Goal-Appraisals, Beliefs, and Psychological Well-Being

The types of coping strategies generated and the types of emotions expressed turn out to be very sensitive indicators of current and future psychological well-being, as measured by depressive mood, positive morale, and measures of self-esteem. Table 16.5 shows the relationship between the proportion of positive emotion state words that occurred during the narration of stressful or traumatic events and the level of depression expressed by the narrator. In all three sets of studies, the data indicate that

TABLE 16.5

Indicators of Well Being Across the Life Span: Proportion of Positive Emotions Expressed When Talking About a Negative, Stressful, or Traumatic Event

Proportion of Positive Emotions Correlated With	*Children*	*Adolescents*	*Adults*
Depressive Mood	–.35	–.38	–.62
Positive Morale Or Self Esteem	+.34	+.32	+.68

Note. All correlations are significant at the $p < .01$ level.

the greater the proportion of reference to positive emotions, the less depressed the narrator was.

The specific linguistic terms the narrators used to refer to emotion (e.g., basic emotion terms as opposed to terms that referred to general positive and negative affective states), the presence of the antecedents and consequences of the specific emotional reaction in narration, and the fact that the emotional event occurred in the past are all determinants of the proportion of positive emotion state words used. These features can be significant predictors of positive morale and depressive mood.

In Table 16.6, we present four different types of appraisals made about emotional, stressful, or traumatic experience. All four of these types of appraisals are significantly related to depressive mood and positive morale. The types of appraisals and beliefs that lead to depressive mood and positive morale in adults are similar to those appraisals that lead to these two mood states in children. The major difference that we find across the life span is not in the appraisals of a goal. Rather, children and adults often differ on *which* goals they consider to be personally meaningful and important. That is, the themes in children's lives are different because the goals they pursue are often different from those of adults. However, when a goal is deemed valuable, children and adults make the same types of appraisals about gains

TABLE 16.6

Appraisals Made About Emotional, Stressful, or Traumatic Experience in Relation to Depressive Mood

	High Depressed			*Low Depressed*		
Appraisals	*Children*	*Adolescents*	*Adults*	*Children*	*Adolescents*	*Adults*
Generated new goals that could be substituted for failed goals	10	23	22	45	58	67
Focused on goals that cannot be attained	53	45	67	23	29	20
Expressed beliefs about lack of ability to cope with the results of loss or aversive states		32	43		06	05
Focused on positive things that were learned as a function of stress and trauma		10	05		35	48

Note. Numbers are expressed as percentages.

and losses. Further, the coping strategies that result in depression are similar over development as are the coping strategies that result in positive well-being.

Table 16.6 shows the percentages of subjects who generated new goals in reaction to having had their goals obstructed in a permanent manner. The percentage of subjects generating new goals was significantly related to whether or not subjects reported high or low degrees of depression, as measured on the Bradburn Affective Balance Scale. The scale was modified for use with young children (Stein & Albro, 1997), but essentially the same items that appear on the adult scale appear on the children's scale. The wording on the children's scale corresponds to children's natural use of emotion terms in everyday interaction (Levine & Stein, 1999). As you can see, the tendency to generate new goals in substitution for goals that have failed is more probable for those subjects who are not highly depressed.

Table 16.6 also shows that significantly more subjects who are high on depressive mood focus more on lost goals than subjects who are low on depressive mood. By focusing on lost goals, we mean that subjects actively fantasize about reinstating the lost goal. We do not mean that subjects are ruminating about past times that were good and meaningful. Even though deeply valued goals are gone forever, many individuals hang on to these goals mentally, and they often have active fantasy sessions that focus on these lost goals.

In addition, Table 16.6 demonstrates that for adolescents and adults, the expression of beliefs about the inability to cope with the results of loss or the inability to muster the personal resources to contend with the devastation of a trauma is significantly related to the level of depression expressed by subjects.

Finally, Table 16.6 shows that a focus on positive outcomes associated with trauma is associated with a lower state of depression. Subjects who were lower on measures of depressive mood were also the ones who mentioned that their stressful or emotional experience, although negative in many respects, had positive benefits and outcomes. The belief that something valuable was learned during a traumatic or stressful event was significantly related to lower states of depression.

CONCLUSIONS

Our emotional and stressful narrative data on children and adults strongly supports the hypothesis that emotional understanding of everyday situations is a rapidly developing skill in early childhood. By 3 years of age, children have a rich array of knowledge about the causes and consequences of basic emotions. The types of strategies children use to cope with emotion are highly similar to the strategies adults use to cope with their emotions.

A strong explanation for the similarities in thinking and coping across the life span is that knowledge about basic emotion is highly constrained to a small number of operating conditions and a small number of coping strategies. Emotional responses occur in reaction to unexpected changes in valued goal states, and even the very young child can monitor whether or not his or her favored goals are being attained or obstructed. The strategies used to cope with changes in a goal's status are also limited to a small number of options. When people fail to attain or maintain goals of value, a decision is first made about the ability to reinstate the goal in question. If people believe that the goal can be reinstated, they activate plans that allow them to proceed with reinstatement. If a valued goal cannot be reinstated, three types of strategies can be activated. The first is to abandon the goal and substitute nothing; the second is to abandon the failed goal and to substitute a new goal for the failed one; the third is to mentally reinstate the lost goal and to fantasize about the conditions that would allow reinstatement, even though the reinstatement cannot occur in reality (Klinger, 1975, 1977, 1987).

The fact that emotional understanding relies on a very limited set of conditions affords young children rapid access to this knowledge. The emotional understanding that children and adults express is then intimately related to decisions about reinstating or substituting new goals. These reinstatement decisions then regulate the experience of both positive and negative mood states. Although adults frequently have more knowledge than children about obstacles to goal attainment and ways of overcoming obstacles, the experience of goal failure and success results in the same types of appraisals and decision processes concerning coping strategies.

An important issue for further research will be to determine the stability of the appraisal and coping strategies activated in emotional situations. A critical issue is whether or not temperamental differences, rather than development differences, regulate the types of appraisals and choice of coping strategies in adverse circumstances. A second issue is whether subjects can learn to change their appraisals and coping strategies.

What has become apparent to us, however, in analyzing the narratives of everyday emotional experience is an extremely productive strategy for describing beliefs, desires, emotions, and plans of people engaged in an ongoing emotion situation. The identification of subjective mental states such as beliefs, desires, emotions, and goals is critical because these states not only predict strategic action that is carried out, but they also predict the type of mood state that will be experienced as a result of the emotional experience.

ACKNOWLEDGMENTS

This research was funded by grants from the Smart Foundation on Early Learning to Nancy L. Stein, by National Institute of Child Health and Human Development Grant HD 25742 to Tom Trabasso and Nancy L. Stein, by a Provost's Grant to Nancy Stein and Tom Trabasso, by a Social Sciences Grant to Nancy Stein and Tom Trabasso, and by National Institute of Mental Health Grants MH44045, MH49985, and MH52517 to Susan Folkman. We want to thank Tom Trabasso and Liz Albro for their valuable conceptual suggestions throughout the writing of this chapter.

REFERENCES

Albro, E. R., & Stein, N. L. (1997). *Children's memory for conflict as a function of liking.* Unpublished manuscript, University of Chicago, Chicago.

Alessandri, S. M., Sullivan, M. W., & Lewis, M. (1990). Violation of expectancy and frustration in early infancy. *Developmental Psychology, 26*(5), 738–744.

Beck, A. T. (1988). *Love is never enough.* New York: Harper & Row.

Folkman, S., & Stein, N. (1997). The analysis of belief and goal processes during reports of stressful events by caregivers of men with AIDS. In N. L. Stein, P. A. Ornstein, B. Tversky, & C. Brainerd (Eds.), *Memory for everyday and emotional events* (pp. 113–138). Mahwah, NJ: Lawrence Erlbaum Associates.

Frijda, N. H. (1995). Passions and socially consequential behavior. In R. Kavanaugh, B. Zimmerberg-Glick, & S. Fein (Eds.), *Emotion: A multidisciplinary perspective* (pp. 1–28). Hillsdale, NJ: Lawrence Erlbaum Associates.

Goldman, S. R. (1982). Knowledge systems for realistic goals. *Discourse Processes, 5,* 279–303.

Gottman, J. (1994). *Why marriages succeed or fail.* New York: Simon & Schuster.

Haith, M. M. (1994). Visual expectations as the first step toward the development of future-oriented processes. In M. M. Haith, J. B. Benson, R. J. Roberts & B. F. Pennington (Eds.), *The development of future-oriented processes.* Chicago: The University of Chicago Press.

Izard, C. E., Hemsbree, E. A., & Huebner, R. R. (1987). Infant's emotional expressions to acute pain: Developmental change and stability of individual differences. *Developmental Psychology, 23,* 105–113.

Kihlstrom, J. (in press-a). Exhumed memory. In S. J. Lynn & N. P. Spanos (Eds.), *Truth in memory* (Reprint). New York: Guilford.
Kihlstrom, J. (in press-b). Trauma and memory. In *Consciousness and cognition: Special issue on the recovery of lost childhood memories for traumatic events* (Reprint).
Klinger, E. (1975). Consequence of commitment to and disengagement from incentives. *Psychological Review, 82,* 1–25.
Klinger, E. (1977). *Meaning and void: Inner experience and the incentives in people's lives.* Minneapolis: University of Minnesota Press.
Klinger, E. (1987). Current concerns and disengagement from incentives. In F. Halisch & J. Kuhl (Eds.), *Motivation, intention, and volition* (pp. 337–347). Berlin: Springer-Verlag.
Kolb, B., & Taylor, L. (1990). Neocortical substrates of emotional behavior. In N. L. Stein, B. Leventhal, & T. Trabasso (Eds.), *Psychological and biological approaches to emotion* (pp. 97–114). Hillsdale, NJ: Lawrence Erlbaum Associates.
LeDoux, J. (1996). *The emotional brain.* New York: Simon & Schuster.
Levine, L. J. (1995). Young children's understanding of the causes of anger and sadness. *Child Development, 66,* 697–709.
Levine, L. J. (1996). The anatomy of disappointment: A naturalistic test of appraisal models of sadness, anger, and hope. *Cognition and Emotion, 10,* 337–359.
Levine, L. J. (in press). Reconstructing memory for emotions. *Journal of Experimental Psychology: General.*
Levine, L. J., & Burgess, S. L. (1995, July). *Beyond arousal: Effects of specific emotions on memory.* Paper presented at the meeting of the Society for Applied Research on Memory and Cognition, Vancouver, Canada.
Levine, L., & Stein, N. L. (1997). *"I'm not his brainö: The emergence of young children's understanding of emotions in themselves and their friends.* Unpublished manuscript, University of Chicago, Chicago.
Levine, L., & Stein, N. L. (1999). Young children's understanding of emotions in themselves and others. Unpublished manuscript. University of Chicago.
Levine, L. J., Stein, N. L., & Liwag, M. (1999). Remembering children's emotions: Sources of concordant and discordant accounts between parents and children. *Developmental Psychology, 35,* 210–230.
Lewis, M., Alessandri, S. M., & Sullivan, M. W. (1990). Violation of expectancy, loss of control, and anger expressions in young infants. *Developmental Psychology, 26*(5), 745–751.
Liwag, M. D., & Stein, N. L. (1995). Children's memory for emotional events: The importance of emotion cue retrieval strategies. *Journal of Experimental Child Psychology, 60,* 2–31.
Mandler, G. (1984). *Mind and body: Psychology of emotion and stress.* New York: Norton.
Michenbaum, D. (1994). *A clinical handbook/practical therapist manual for assessing and treating adults with post-traumatic stress disorder.* Waterloo, Ontario, Canada: Institute Press.
Oatley, K., & Johnson-Laird, P. N. (1987). Towards a cognitive theory of emotion. *Cognition and Emotion, 1,* 29–50.
Roseman, I. J., Antoniou, A. A., & Jose, P. E. (1996). Appraisal determinants of emotions: Constructing a more accurate and comprehensive theory. *Cognition and Emotion, 10*(3), 241–277.
Ross, H. (1996). Negotiating principles of entitlement in sibling property disputes. *Developmnetal Psychology, 32,* 90–101.
Ross, M., & Holmberg, D. (1990). Recounting the past. Gender differences in the recall of events in the history of a close relationship. In M. Olson & M. P. Zanna (Eds.),

Self-inferences process: The Ontario Symposium (Vol. 6, pp. 135–152). Hillsdale, NJ: Lawrence Erlbaum Associates.

Smith, C. A., & Lazarus, R. S. (1993). Appraisal components, core relational themes, and the emotions. *Cognition and Emotion, 7*(3/4), 233–269.

Sroufe, A. (1979). Socioemotional development. In J. Osofsky (Ed.), *The handbook of infant development* (pp. 462–516). New York: Wiley.

Stein, N. L. (1984). Critical issues in the development of literacy: Toward a theory of learning and instruction. *American Journal of Education, 93*(1), 177–199.

Stein, N. L. (1988). The development of storytelling skill. In M. B. Franklin & S. Barten (Eds.), *Child language: A book of readings* (pp. 282–297). New York: Cambridge University Press.

Stein, N. L., & Albro, E. R. (1997, April). *Children's understanding of conflict: Evidence from past memories and on-line negotiated resolutions.* Paper presented at the meeting of the Society for Research in Child Development, Washington, DC.

Stein, N.L. & Bernas, R. (1999). The early emergence of argumentative knowledge and skills. In J. Andriessen & P. Coirier (Eds.), *Foundations of argumentative text processing.* Amsterdam: Amsterdam University Press.

Stein, N. L., Bernas, R., & Calicchia, D. (1997). Conflict talk: Understanding and resolving arguments. In T. Givon (Ed.), *Typological studies in language: Conversational analysis* (pp. 233–267). Amsterdam: John Benjamins.

Stein, N. L., & Boyce, W. T. (1995, April). *The role of physiological reactivity in attending to, remembering, and responding to an emotional event.* Paper presented at the meeting of the Society for Research in Child Development, Indianapolis.

Stein, N. L., & Chae, D. (1999). *Predicting recovery from the death of a partner.* Manuscript in preparation.

Stein, N. L., Folkman, S., Trabasso, T., & Richards, T. A. (1997). Appraisal and goal processes as predictors of psychological well-being in bereaved caregivers. *Journal of Personality and Social Psychology, 72*(4), 872–884.

Stein, N. L., & Jewett, J. (1986). A conceptual analysis of the meaning of basic negative emotions: Implications for a theory of development. In C. E. Izard & P. Read (Eds.), *Measurement of emotion in infants and children, vol. 2* (pp. 238–268). New York: Cambridge University Press.

Stein, N. L., & Levine, L. J. (1987). Thinking about feelings: The development and organization of emotional knowledge. In R. E. Snow & M. Farr (Eds.), *Aptitude, learning, and instruction: Vol. 3. Cognition, conation and affect* (pp. 168–198). Hillsdale, NJ: Lawrence Erlbaum Associates.

Stein, N. L., & Levine, L. J. (1989). The causal organization of emotional knowledge. *Cognition and Emotion, 3,* 343–378.

Stein, N. L., & Levine, L. J. (1990). Making sense out of emotion: The representation and use of goal-structured knowledge. In N. L. Stein, B. Leventhal, & T. Trabasso (Eds.), *Psychological and biological approaches to emotion* (pp. 45–74). Hillsdale, NJ: Lawrence Erlbaum Associates.

Stein, N. L., & Liwag, M. D. (1997). A goal-appraisal process approach to understanding and remembering emotional events. In P. van den Broek, P. Bauer, & T. Bourg (Eds.), *Developmental spans in event comprehension and representation* (pp. 199–236). Mahwah, NJ: Lawrence Erlbaum Associates.

Stein, N. L., Liwag, M. D., & Wade, E. (1995). A goal-based approach to memory for emotional events: Implications for theories of understanding and socialization. In R.

Kavanaugh, B. Zimmerberg-Glick, & S. Fein (Eds.), *Emotion: A multidisciplinary perspective* (pp. 91–118). Hillsdale, NJ: Lawrence Erlbaum Associates.

Stein, N. L., & Ross, M. (1996, November). *Remembering arguments: Similarities and differences in husband's and wife's description of the same conflicts.* Paper presented at the meeting of the Psychonomic Society, Chicago.

Stein, N. L., Ross, M. A., Sheldrick, R., & Fergusson, P. (1998). Writing about trauma from two different perspectives: Effects on narration and psychological well-being.

Stein, N. L., & Trabasso, T. (1989). Children's understanding of changing emotion states. In C. Saarni & P. Harris (Eds.), *The development of emotional understanding* (pp. 50–77). New York: Cambridge University Press.

Stein, N. L., Trabasso, T. (1982). What's in a story: An approach to comprehension and instruction. In R. Glaser (Ed.), *Advances in instructional psychology* (Vol. 2, pp. 212–267). Hillsdale, NJ: Lawrence Erlbaum Associates.

Stein, N. L., & Trabasso, T. (1992). The organization of emotional experience: Creating links among emotion, thinking, language and intentional action [Special issue]. *Cognition and Emotion, 6*(3/4).

Stein, N. L., & Trabasso, T. (1999). *Can we experience positive mood states at the same time we experience depression: Unraveling the relationship between depressive moods and positive morale.*

Stein, N. L., Trabasso, T., & Albro, E. R. (1997). *The course of caregiving in an ever-worsening illness.* Unpublished manuscript, University of Chicago, Chicago.

Stein, N. L., Trabasso, T., & Lichtenthal, M. (1999). *Predicting well-being from reactions to stressful events during caregiving experiences.*

Stein, N. L., Trabasso, T., & Liwag, M. (1992). The representation and organization of emotional experience: Unfolding the emotion episode. In M. Lewis & J. Haviland (Eds.), *Handbook of emotions* (pp. 279–300) New York: Guilford.

Stein, N. L., Trabasso, T., & Liwag, M. D. (1994). The Rashomon phenomenon: Personal frames and future-oriented appraisals in memory for emotional events. In M. Haith, J. Benson, R. Roberts, & B. Pennington (Eds.), *The development of future-oriented processes* (pp. 323–349). Chicago: University of Chicago Press.

Stein, N. L., Wade, E., & Liwag, M. D. (1997). A theoretical approach to understanding and remembering emotional events. In N. L. Stein, P. A. Ornstein, B. Tversky, & C. Brainerd (Eds.), *Memory for everyday and emotional events* (pp. 15–48). Mahwah, NJ: Lawrence Erlbaum Associates.

Stenberg, C. R., & Campos, J. J. (1990). The development of anger expressions in infancy. In N. L. Stein, B. Leventhal, & T. Trabasso (Eds.), *Psychological and biological approaches to emotion* (pp. 247–282). Hillsdale, NJ: Lawrence Erlbaum Associates.

Stern, D. (1985). *The interpersonal world of the infant.* New York: Basic Books.

Stern, D. (1992). *Diary of a baby.* New York: Basic Books.

Trabasso, T. & Ozyureck, A. (1997). Communicating evaluation in narrative understanding. In T. Givon (Ed.), *Conversation: Cognitive, communicative, and social perspectives* (pp. 269–302). Amsterdam: John Benjamins Publishing Company.

Trabasso, T., & Stein, N. L. (1994). Using goal-plan knowledge to merge the past with the present and the future in narrating events on-line. In M. Haith, J. Benson, R. Roberts, & B. Pennington (Eds.), *The development of future-oriented processes* (pp. 323–349). Chicago: University of Chicago Press.

Trabasso, T., & Stein, N. L. (1997). Narrating, representing, and remembering event sequences. In P. van den Broek, P. Bauer & T. Bourg (Eds.), *Developmental spans in event comprehension and representation.* Mahwah, NJ: Lawrence Erlbaum Associates.

Trabasso, T., Stein, N. L., & Johnson, L. R. (1981). Children's knowledge of events: A causal analysis of story structure. In G. H. Bower (Ed.), *Learning and motivation* (Vol. 15, pp. 237–282). New York: Academic Press.

Traue, H. C., & Pennebaker, J. W. (1994). *Emotion, inhibition, and health.* Gottingen, Germany: Hogrefe & Huber.

Author Index

Subject Index